T0246408

Crop Protection and Management

Crop Protection and Management

Edited by
Frazer Murphy

Larsen & Keller
www.larsen-keller.com

Crop Protection and Management
Edited by Frazer Murphy
ISBN: 978-1-63549-079-4 (Hardback)

© 2017 Larsen & Keller

 Larsen & Keller

Published by Larsen and Keller Education,
5 Penn Plaza,
19th Floor,
New York, NY 10001, USA

Cataloging-in-Publication Data

Crop protection and management / edited by Frazer Murphy.
    p. cm.
Includes bibliographical references and index.
ISBN 978-1-63549-079-4
1. Plants, Protection of. 2. Crops. 3. Cropping systems.
I. Murphy, Frazer.
SB950 .C76 2017
632.9--dc23

This book contains information obtained from authentic and highly regarded sources. All chapters are published with permission under the Creative Commons Attribution Share Alike License or equivalent. A wide variety of references are listed. Permissions and sources are indicated; for detailed attributions, please refer to the permissions page. Reasonable efforts have been made to publish reliable data and information, but the authors, editors and publisher cannot assume any responsibility for the vailidity of all materials or the consequences of their use.

Trademark Notice: All trademarks used herein are the property of their respective owners. The use of any trademark in this text does not vest in the author or publisher any trademark ownership rights in such trademarks, nor does the use of such trademarks imply any affiliation with or endorsement of this book by such owners.

The publisher's policy is to use permanent paper from mills that operate a sustainable forestry policy. Furthermore, the publisher ensures that the text paper and cover boards used have met acceptable environmental accreditation standards.

Printed and bound in the United States of America.

For more information regarding Larsen and Keller Education and its products, please visit the publisher's website www.larsen-keller.com

Table of Contents

Preface

Crop management is the practice of using various techniques and methods to increase crop yield, better crop quality, improve growth and development of crops. Different crops require different managing techniques, determined by their biological characteristics and climatic requirements. This book includes detailed information about the fundamental principles and practices that can be used for crop protection. It provides the readers with extensive material to understand the concepts and applications of crop management. It aims to serve as a resource guide for graduates and post-graduates alike and contribute to the growth of the discipline.

A foreword of all Chapters of the book is provided below:

Chapter 1 - From the early days of civilization when man gave up on a nomadic life and began settling along the banks of rivers, the cultivation of certain plants for food gained popularity. Crops can be divided into categories based on the reasons behind their cultivation, the seasons of cultivation and so on. This chapter introduces the reader to the science of agriculture and gives a brief history about the cultivation of crops as well as the types of crops; **Chapter 2** - The study of diseases in plants caused by pathogens and certain environmental factors is called plant pathology. This chapter examines in minutiae the common pathogens, the diseases and the infection methods used by the pathogens to spread disease in the crop. It also examines the natural and man-made causes of diseases in plants; **Chapter 3** - Crop yield refers to the seed generation by the plant and also the output per unit area of cultivated land. It is the measure that helps determine the yield potential of strains of cultivated crop which ultimately helps farmers and agricultural scientists determine the favorable conditions for higher output and also to establish better hybrid varieties of the crop. This chapter sheds light on the importance of crop yielding and gaps in yielding as well; **Chapter 4** - The main concern of crop management is to maximize agricultural output while reducing strain on the cultivated land. With this in mind a multitude of cropping methods and practices have been developed and this chapter explains in detail methods like intercropping, monocropping, sharecropping, multicropping and pollination management listing the pros and cons of each; **Chapter 5** - Apart from hybridization of crops, another obvious method of increasing crop output is by the use of pesticides and insecticides to minimize the loss in crop that occurs due to pests and insects. This chapter elucidates the various methods employed for this purpose- spraying of pesticides and insecticides, bird netting, biological alternative to manmade pesticides and the like. Each of these methods has been explained in detail citing examples wherever needed; **Chapter 6** - Harvest is the final stage of cropping and plays a vital role in determining the profit potential of the cultivated crop. Post-harvest refers to that stage when the grain or produce is harvested from the plant; the main concern at this stage is to reduce spoilage, wastage and to ensure that the harvest is protected from pests, disease and decay. This chapter details the methods used to increase the post-harvest shelf life of the produce and the causes of post-harvest losses of grains and vegetables.

I would like to thank the entire editorial team who made sincere efforts for this book and my family who supported me in my efforts of working on this book. I take this opportunity to thank all those who have been a guiding force throughout my life.

Editor

Introduction to Crop

From the early days of civilization when man gave up on a nomadic life and began settling along the banks of rivers, the cultivation of certain plants for food gained popularity. Crops can be divided into categories based on the reasons behind their cultivation, the seasons of cultivation and so on. This chapter introduces the reader to the science of agriculture and gives a brief history about the cultivation of crops as well as the types of crops.

Crop

A crop is any cultivated plant, fungus, or alga that is harvested for food, clothing, livestock fodder, biofuel, medicine, or other uses. In contrast, animals that are raised by humans are called livestock, except those that are kept as pets. Microbes, such as bacteria or viruses, are referred to as cultures. Microbes are not typically grown for food, but are rather used to alter food. For example, bacteria are used to ferment milk to produce yogurt.

Crops drying in a home in Punjab, India.

Major crops include sugarcane, pumpkin, maize (corn), wheat, rice, cassava, soybeans, hay, potatoes and cotton.

Based on the growing season, the crops grown in India can be classified as kharif crops and rabi crops.

Domesticated plants

Bumper Crop

In agriculture, a bumper crop is a crop that has yielded an unusually productive harvest.

A bumper crop can also be a source of problems, such as when there is insufficient storage space (barns, grain bins, etc.) for an overlarge crop. The word "bumper" has a second definition meaning "something unusually large," which is where this term comes from.

Related Meanings or Puns

The term "bumper crop" has also been used to refer to a similar large result in other activities, or as a pun such as with a group of automobiles (for their front/rear bumpers).

Kharif Crop

Kharif crops or monsoon crops are domesticated plants cultivated and harvested during the rainy (monsoon) season in the South Asia, which lasts between April and October depending on the area. Main kharif crops are millet and rice.

Kharif Season

Kharif crops are usually sown with the beginning of the first rains in July, during the south-west monsoon season. In Pakistan the kharif season starts on April 16th and lasts until October 15th. In India the kharif season varies by crop and state, with kharif starting at the earliest in May and ending at the latest in January, but is popularly considered to start in June and to end in October. Kharif stand in contrast with the Rabi crops, cultivated during the dry season. Both words came with the arrival of Mughals in the Indian subcontinent and are widely used ever-since. *Kharif* means "autumn" in Arabic. Since this period coincides with the beginning of autumn / winter in the Indian sub-continent, it is called "Kharif period".

Kharif crops are usually sown with the beginning of the first rains towards the end of May in the southern state of Kerala during the advent of south-west monsoon season. As the monsoon rains advance towards the north India, the sowing dates vary accordingly and reach July in north Indian states.

These crops are dependent on the quantity of rain water as well its timing. Too much, too little or at wrong time may lay waste the whole year's efforts.

Common Kharif Crops

- Rice (paddy and deepwater)
- Millet
- Maize (corn)
- Mung bean (green gram)
- Urad bean (black gram)
- Guar
- Pea
- Peanut (groundnut)

Examples of Kharif Crop

Rice

A mixture of brown, white, and red indica rice, also containing wild rice, *Zizania* species

Rice is the seed of the grass species *Oryza sativa* (Asian rice) or *Oryza glaberrima* (African rice). As a cereal grain, it is the most widely consumed staple food for a large part of the world's human population, especially in Asia. It is the agricultural commodity with the third-highest worldwide production, after sugarcane and maize, according to 2012 FAOSTAT data.

Oryza sativa with small wind-pollinated flowers

Since a large portion of maize crops are grown for purposes other than human consumption, rice is the most important grain with regard to human nutrition and caloric intake, providing more than one-fifth of the calories consumed worldwide by humans.

Cooked brown rice from Bhutan

Wild rice, from which the crop was developed, may have its native range in Australia. Chinese legends attribute the domestication of rice to Shennong, the legendary emperor of China and inventor of Chinese agriculture. Genetic evidence has shown that rice originates from a single domestication 8,200–13,500 years ago in the Pearl River valley region of China. Previously, archaeological evidence had suggested that rice was domesticated in the Yangtze River valley region in China.

From East Asia, rice was spread to Southeast and South Asia. Rice was introduced to Europe through Western Asia, and to the Americas through European colonization.

There are many varieties of rice and culinary preferences tend to vary regionally. In some areas such as the Far East or Spain, there is a preference for softer and stickier varieties.

Rice can come in many shapes, colours and sizes. Photo by the IRRI.

Rice, a monocot, is normally grown as an annual plant, although in tropical areas it can survive as a perennial and can produce a ratoon crop for up to 30 years. The rice plant can grow to 1–1.8 m (3.3–5.9 ft) tall, occasionally more depending on the variety and soil fertility. It has long, slender leaves 50–100 cm (20–39 in) long and 2–2.5 cm (0.79–0.98 in) broad. The small wind-pollinated flowers are produced in a branched arching to pendulous inflorescence 30–50 cm (12–20 in) long. The edible seed is a grain (caryopsis) 5–12 mm (0.20–0.47 in) long and 2–3 mm (0.079–0.118 in) thick.

Oryza sativa, commonly known as Asian rice

Rice cultivation is well-suited to countries and regions with low labor costs and high rainfall, as it is labor-intensive to cultivate and requires ample water. However, rice can be grown practically anywhere, even on a steep hill or mountain area with the use of water-controlling terrace systems. Although its parent species are native to Asia and certain parts of Africa, centuries of trade and exportation have made it commonplace in many cultures worldwide.

The traditional method for cultivating rice is flooding the fields while, or after, setting the young seedlings. This simple method requires sound planning and servicing of the water damming and channeling, but reduces the growth of less robust weed and pest plants that have no submerged growth state, and deters vermin. While flooding is not mandatory for the cultivation of rice, all other methods of irrigation require higher effort in weed and pest control during growth periods and a different approach for fertilizing the soil.

The name wild rice is usually used for species of the genera *Zizania* and *Porteresia*, both wild and domesticated, although the term may also be used for primitive or uncultivated varieties of *Oryza*.

Etymology

First used in English in the middle of the 13th century, the word "rice" derives from the Old French *ris*, which comes from Italian *riso*, in turn from the Latin *oriza*, which derives from the Greek (*oruza*). The Greek word is the source of all European words (cf. Welsh *reis*, German *Reis*, Lithuanian *ryžiai*, Serbo-Croatian *riža*, Polish *ryż*, Dutch *rijst*, Hungarian *rizs*, Romanian *orez*).

The origin of the Greek word is unclear. It is sometimes held to be from the Tamil word அரிசி (*arisi*), or rather Old Tamil *arici*. However, Krishnamurti disagrees with the notion that Old Tamil *arici* is the source of the Greek term, and proposes that it was borrowed from descendants of Proto-Dravidian *wariñci* instead. Mayrhofer suggests that the immediate source of the Greek word is to be sought in Old Iranian words of the types *vrīz- or *vrinj-, but these are ultimately traced back to Indo-Aryan (as in Sanskrit *vrīhí-*) and subsequently to Dravidian by Witzel and others.

Cooking

The varieties of rice are typically classified as long-, medium-, and short-grained. The grains of long-grain rice (high in amylose) tend to remain intact after cooking; medium-grain rice (high in amylopectin) becomes more sticky. Medium-grain rice is used for sweet dishes, for *risotto* in Italy, and many rice dishes, such as *arròs negre*, in Spain. Some varieties of long-grain rice that are high in amylopectin, known as Thai Sticky rice, are usually steamed. A stickier medium-grain rice is used for *sushi*; the stickiness allows rice to hold its shape when molded. Short-grain rice is often used for rice pudding.

Instant rice differs from parboiled rice in that it is fully cooked and then dried, though there is a significant degradation in taste and texture. Rice flour and starch often are used in batters and breadings to increase crispiness.

Preparation

Rice is typically rinsed before cooking to remove excess starch. Rice produced in the US is usually fortified with vitamins and minerals, and rinsing will result in a loss of nutrients. Rice may be rinsed repeatedly until the rinse water is clear to improve the texture and taste.

Rice may be soaked to decrease cooking time, conserve fuel, minimize exposure to high temperature, and reduce stickiness. For some varieties, soaking improves the texture of the cooked rice by increasing expansion of the grains. Rice may be soaked for 30 minutes up to several hours.

Milled to unmilled rice, from left to right, white rice *(Japanese rice)*, rice with germ, brown rice

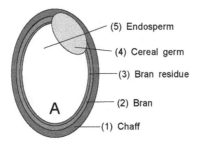

A: Rice with chaff

B: Brown rice

C: Rice with germ

D: White rice with bran residue

E: Musenmai (Japanese: 無洗米), "Polished and ready to boil rice", literally, non-wash rice

(1): Chaff

(2): Bran

(3): Bran residue

(4): Cereal germ

(5): Endosperm

Brown rice may be soaked in warm water for 20 hours to stimulate germination. This process, called germinated brown rice (GBR), activates enzymes and enhances amino acids including gamma-aminobutyric acid to improve the nutritional value of brown rice. This method is a result of research carried out for the United Nations International Year of Rice.

Processing

Rice is cooked by boiling or steaming, and absorbs water during cooking. With the absorption method, rice may be cooked in a volume of water similar to the volume of rice. With the rapid-boil method, rice may be cooked in a large quantity of water which is drained before serving. Rapid-boil preparation is not desirable with enriched rice, as much of the enrichment additives are lost when the water is discarded. Electric rice cookers, popular in Asia and Latin America, simplify the process of cooking rice. Rice (or any other grain) is sometimes quickly fried in oil or fat before boiling (for example saffron rice or risotto); this makes the cooked rice less sticky, and is a cooking style commonly called pilaf in Iran and Afghanistan or biryani (Dam-pukhtak) in India and Pakistan.

Dishes

In Arab cuisine, rice is an ingredient of many soups and dishes with fish, poultry, and other types of meat. It is also used to stuff vegetables or is wrapped in grape leaves (dolma). When combined with milk, sugar, and honey, it is used to make desserts. In some regions, such as Tabaristan, bread is made using rice flour. Medieval Islamic texts spoke of medical uses for the plant. Rice may also be made into congee (also called rice porridge, fawrclaab, okayu, Xifan, jook, or rice gruel) by adding more water than usual, so that the cooked rice is saturated with water, usually to the point that it disintegrates. Rice porridge is commonly eaten as a breakfast food, and is also a traditional food for the sick.

Nutrition and Health

Importance

Rice is the staple food of over half the world's population. It is the predominant dietary energy source for 17 countries in Asia and the Pacific, 9 countries in North and South America and 8 countries in Africa. Rice provides 20% of the world's dietary energy supply, while wheat supplies 19% and maize (corn) 5%.

A detailed analysis of nutrient content of rice suggests that the nutrition value of rice varies based on a number of factors. It depends on the strain of rice, that is between white, brown, red, and black (or purple) varieties of rice – each prevalent in different parts of the world. It also depends on nutrient quality of the soil rice is grown in, whether and how the rice is polished or processed, the manner it is enriched, and how it is prepared before consumption.

An illustrative comparison between white and brown rice of protein quality, mineral and vitamin quality, carbohydrate and fat quality suggests that neither is a complete nutrition source. Between the two, there is a significant difference in fiber content and minor differences in other nutrients.

Comparison to other Major Staple Foods

Nutrient contents in %DV of common foods (raw, uncooked) per 100 g

Food	Protein		Fiber	Vitamins													Minerals									
	DV	Q	DV	A	B1	B2	B3	B5	B6	B9	B12	Ch.	C	D	E	K	Ca	Fe	Mg	P	K	Na	Zn	Cu	Mn	Se
cooking Reduction %				10	30	20	25		25	35	0	0	30				10	15	20	10	20	5	10	25		
Corn	20	55	6	1	13	4	16	4	19	19	0	0	0	0	0	1	1	11	31	34	15	1	20	10	42	0
Rice	14	71	1.3	0	12	3	11	20	5	2	0	0	0	0	0	0	1	9	6	7	2	0	8	9	49	22
Wheat	27	51	40	0	28	7	34	19	21	11	0	0	0	0	0	0	3	20	36	51	12	0	28	28	151	128
Soybean	73	132	0	31	58	51	8	8	19	94	0	24	10	0	4	59	28	87	70	70	51	0	33	83	126	25
Pigeon pea	43	91	1	50	43	11	15	13	13	114	0	0	0	0	0	0	13	29	46	37	40	1	18	53	90	12
Potato	4	112	7.3	0	5	2	5	3	15	4	0	0	33	0	0	2	1	4	6	6	12	0	2	5	8	0
Sweet potato	3	82	10	284	5	4	3	8	10	3	0	0	4	0	1	2	3	3	6	5	10	2	2	8	13	1
Spinach	6	119	7.3	188	5	11	4	1	10	49	0	4.5	47	0	10	604	10	15	20	5	16	3	4	6	45	1
Dill	7	32	7	154	4	17	8	4	9	38	0	0	142	0	0	0	21	37	14	7	21	3	6	7	63	0
Carrots	2	24	9.3	334	4	3	5	3	7	5	0	0	10	0	3	16	3	2	3	4	9	3	2	2	7	0
Guava	5	24	18	12	4	2	5	5	6	12	0	0	381	0	4	3	2	1	5	4	12	0	2	11	8	1
Papaya	1	7	5.6	22	2	2	2	2	1	10	0	0	103	0	4	3	2	1	2	1	7	0	0	1	1	1
Pumpkin	2	56	1.6	184	3	6	3	3	3	4	0	0	15	0	5	1	2	4	3	4	10	0	2	6	6	0
Sunflower oil	0		0	0	0	0	0	0	0	0	0	0	0	0	205	7	0	0	0	0	0	0	0	0	0	0
Egg	25	136	0	10	5	28	0	14	7	12	22	45	0	9	5	0	5	10	3	19	4	6	7	5	2	45
Milk	6	138	0	2	3	11	1	4	2	1	7	2.6	0	0	0	0	11	0	2	9	4	2	3	1	0	5

Ch. = Choline; Ca = Calcium; Fe = Iron; Mg = Magnesium; P = Phosphorus; K = Potassium; Na = Sodium; Zn = Zinc; Cu = Copper; Mn = Manganese; Se = Selenium; %DV = % daily value i.e. % of DRI (Dietary Reference Intake) Note: All nutrient values including protein and fiber are in %DV per 100 grams of the food item. Significant values are highlighted in light Gray color and bold letters. Cooking reduction = % Maximum typical reduction in nutrients due to boiling without draining for ovo-lacto-vegetables group Q = Quality of Protein in terms of completeness without adjusting for digestability.

The table below shows the nutrient content of major staple foods in a raw form. Raw grains, however, are not edible and can not be digested. These must be sprouted, or prepared and cooked for human consumption. In sprouted and cooked form, the relative nutritional and anti-nutritional contents of each of these grains is remarkably different from that of raw form of these grains reported in this table.

Nutrient content of major staple foods												
STAPLE:	RDA	Maize / Corn[A]	Rice (white) [B]	Rice (brown) [I]	Wheat[C]	Potato[D]	Cassava[E]	Soybean (Green) [F]	Sweet potato[G]	Soghum[H]	Yam[Y]	Plantain[Z]
Component (per 100g portion)	Amount	Amount	Amount	Amount	Amount	Amount	Amount	Amount	Amount	Amount	Amount	Amount
Water (g)	3000	10	12	10	13	**79**	60	68	77	9	70	65
Energy (kJ)		1528	1528	**1549**	1369	322	670	615	360	1419	494	511
Protein (g)	50	9.4	7.1	7.9	12.6	2.0	1.4	**13.0**	1.6	11.3	1.5	1.3
Fat (g)		4.74	0.66	2.92	1.54	0.09	0.28	**6.8**	0.05	3.3	0.17	0.37
Carbohydrates (g)	130	74	**80**	77	71	17	38	11	20	75	28	32
Fiber (g)	30	7.3	1.3	3.5	**12.2**	2.2	1.8	4.2	3	6.3	4.1	2.3
Sugar (g)		0.64	0.12	0.85	0.41	0.78	1.7	0	4.18	0	0.5	**15**
Calcium (mg)	1000	7	28	23	29	12	16	**197**	30	28	17	3
Iron (mg)	8	2.71	0.8	1.47	3.19	0.78	0.27	3.55	0.61	**4.4**	0.54	0.6
Magnesium (mg)	400	127	25	**143**	126	23	21	65	25	0	21	37
Phosphorus (mg)	700	210	115	**333**	288	57	27	194	47	287	55	34
Potassium (mg)	4700	287	115	223	363	421	271	620	337	350	**816**	499
Sodium (mg)	1500	35	5	7	2	6	14	15	**55**	6	9	4
Zinc (mg)	11	2.21	1.09	2.02	**2.65**	0.29	0.34	0.99	0.3	0	0.24	0.14
Copper (mg)	0.9	0.31	0.22		**0.43**	0.11	0.10	0.13	0.15	-	0.18	0.08

Manganese (mg)	2.3	0.49	1.09	3.74	**3.99**	0.15	0.38	0.55	0.26	-	0.40	-
Selenium (µg)	55	15.5	15.1		**70.7**	0.3	0.7	1.5	0.6	0	0.7	1.5
Vitamin C (mg)	90	0	0	0	0	19.7	20.6	**29**	2.4	0	17.1	18.4
Thiamin (B1)(mg)	1.2	0.39	0.07	0.40	0.30	0.08	0.09	**0.44**	0.08	0.24	0.11	0.05
Riboflavin (B2)(mg)	1.3	**0.20**	0.05	0.09	0.12	0.03	0.05	0.18	0.06	0.14	0.03	0.05
Niacin (B3) (mg)	16	3.63	1.6	5.09	**5.46**	1.05	0.85	1.65	0.56	2.93	0.55	0.69
Pantothenic acid (B5 (mg)	5	0.42	1.01	**1.49**	0.95	0.30	0.11	0.15	0.80	-	0.31	0.26
Vitamin B6 (mg)	1.3	**0.62**	0.16	0.51	0.3	0.30	0.09	0.07	0.21	-	0.29	0.30
Folate Total (B9) (µg)	400	19	8	20	38	16	27	**165**	11	0	23	22
Vitamin A (IU)	5000	214	0	0	9	2	13	180	**14187**	0	138	1127
Vitamin E, alpha-tocopherol (mg)	15	0.49	0.11	0.59	**1.01**	0.01	0.19	0	0.26	0	0.39	0.14
Vitamin K1 (µg)	120	0.3	0.1	1.9	1.9	1.9	1.9	0	1.8	0	**2.6**	0.7
Beta-carotene (µg)	10500	97	0		5	1	8	0	**8509**	0	83	457
Lutein+zeaxanthin (µg)		**1355**	0		220	8	0	0	0	0	0	30
Saturated fatty acids (g)		0.67	0.18	0.58	0.26	0.03	0.07	**0.79**	0.02	0.46	0.04	0.14
Monounsaturated fatty acids (g)		1.25	0.21	1.05	0.2	0.00	0.08	**1.28**	0.00	0.99	0.01	0.03
Polyunsaturated fatty acids (g)		2.16	0.18	1.04	0.63	0.04	0.05	**3.20**	0.01	1.37	0.08	0.07

A corn, yellow	B rice, white, long-grain, regular, raw, unenriched
C wheat, hard red winter	D potato, flesh and skin, raw
E cassava, raw	F soybeans, green, raw
G sweet potato, raw, unprepared	H sorghum, raw
Y yam, raw	Z plantains, raw
I rice, brown, long-grain, raw	

Arsenic Concerns

Rice and rice products contain arsenic, a known poison and Group 1 carcinogen. There is no safe level of arsenic, but, as of 2012, a limit of 10 parts per billion has been established in the United States for drinking water, twice the level of 5 parts per billion originally proposed by the EPA. Consumption of one serving of some varieties of rice gives more exposure to arsenic than consumption of 1 liter of water that contains 5 parts per billion arsenic; however, the amount of arsenic in rice varies widely with the greatest concentration in brown rice and rice grown on land formerly used to grow cotton; in the United States, Arkansas, Louisiana, Missouri, and Texas. The U.S. Food and Drug Administration (FDA) is studying this issue, but has not established a limit. China has set a limit of 150 ppb for arsenic in rice.

White rice grown in Arkansas, Louisiana, Missouri, and Texas, which account for 76 percent of American-produced rice had higher levels of arsenic than other regions of the world studied, possibly because of past use of arsenic-based pesticides to control cotton weevils. Jasmine rice from Thailand and Basmati rice from Pakistan and India contain the least arsenic among rice varieties in one study.

Bacillus Cereus

Cooked rice can contain *Bacillus cereus* spores, which produce an emetic toxin when left at 4–60 °C (39–140 °F). When storing cooked rice for use the next day, rapid cooling is advised to reduce the risk of toxin production. One of the enterotoxins produced by *Bacillus cereus* is heat-resistant; reheating contaminated rice kills the bacteria, but does not destroy the toxin already present.

Rice-growing Environments

Rice can be grown in different environments, depending upon water availability. Generally, rice does not thrive in a waterlogged area, yet it can survive and grow herein and it can also survive flooding.

1. Lowland, rainfed, which is drought prone, favors medium depth; waterlogged, submergence, and flood prone

2. Lowland, irrigated, grown in both the wet season and the dry season

3. Deep water or floating rice

4. Coastal Wetland

5. Upland rice is also known as Ghaiya rice, well known for its drought tolerance

History of Domestication and Cultivation

There have been plenty of debates on the origins of the domesticated rice. Genetic evidence published in the *Proceedings of the National Academy of Sciences of the United States of America* (PNAS) shows that all forms of Asian rice, both *indica* and *japonica*, spring from a single domestication that occurred 8,200–13,500 years ago in China of the wild rice *Oryza rufipogon*. A 2012 study published in *Nature*, through a map of rice genome variation, indicated that the domestication of rice occurred in the Pearl River valley region of China based on the genetic evidence. From East Asia, rice was spread to South and Southeast Asia. Before this research, the commonly accepted view, based on archaeological evidence, is that rice was first domesticated in the region of the Yangtze River valley in China.

Rice broker in 1820s Japan of the Edo period (*"36 Views of Mount Fuji"* Hokusai)

Morphological studies of rice phytoliths from the Diaotonghuan archaeological site clearly show the transition from the collection of wild rice to the cultivation of domesticated rice. The large number of wild rice phytoliths at the Diaotonghuan level dating from 12,000–11,000 BP indicates that wild rice collection was part of the local means of subsistence. Changes in the morphology of Diaotonghuan phytoliths dating from 10,000–8,000 BP show that rice had by this time been domesticated. Soon afterwards the two major varieties of indica and japonica rice were being grown in Central China. In the late 3rd millennium BC, there was a rapid expansion of rice cultivation into mainland Southeast Asia and westwards across India and Nepal.

In 2003, Korean archaeologists claimed to have discovered the world's oldest domesticated rice. Their 15,000-year-old age challenges the accepted view that rice cultivation originated in China about 12,000 years ago. These findings were received by academia with strong skepticism, and the results and their publicizing has been cited as being driven by a combination of nationalist and regional interests. In 2011, a combined effort by the Stanford University, New York University, Washington University in St. Louis, and Purdue University has provided the strongest evidence yet that there is only one single origin of domesticated rice, in the Yangtze Valley of China.

Rice spread to the Middle East where, according to Zohary and Hopf (2000, p. 91), *O. sativa* was recovered from a grave at Susa in Iran (dated to the 1st century AD).

Regional History

In a recent study, scientist have found a link for differences in human culture based on either wheat or rice cultivating races since ancient times.

Africa

Rice crop in Madagascar

African rice has been cultivated for 3500 years. Between 1500 and 800 BC, *Oryza glaberrima* propagated from its original centre, the Niger River delta, and extended to Senegal. However, it never developed far from its original region. Its cultivation even declined in favour of the Asian species, which was introduced to East Africa early in the common era and spread westward. African rice helped Africa conquer its famine of 1203.

Asia

Ricefields at Santa Maria, Bulacan, Philippines

Rice fields in Dili/East Timor

Today, the majority of all rice produced comes from China, India, Indonesia, Bangladesh, Vietnam, Thailand, Myanmar, Pakistan, Philippines, Korea and Japan. Asian farmers still account for 87% of the world's total rice production.

Indian women separating rice from straw

Cambodian women planting rice.

Nepal

Rice is the major food amongst all the ethnic groups in Nepal. Agriculture in Madesh mainly depends on the rice cultivation during rainy season in trai areas of Nepal. Rice production is acutely dependent on rainfall and farmers use irrigation channels throughout the cultivation seasons with to the support of the Government and NNF Nepal . The principal cultivation season, known as "Berna-Bue Charne", is from June to July and the subsidiary cultivation season, known as "Ropai, is from April to September. During Ropai period, there is usually enough water to sustain the cultivation of all rice fields, nevertheless in Berna-Bue Charne period, there is only enough water for cultivation of few of the land extent. The Agricultural Development Office of every district take care of crop in Nepal.

Philippines

The Banaue Rice Terraces (Filipino: *Hagdan-hagdang Palayan ng Banawe*) are 2,000-year-old terraces that were carved into the mountains of Ifugao in the Philippines by ancestors of the indigenous people. The Rice Terraces are commonly referred to as the "Eighth Wonder of the World". It is commonly thought that the terraces were built with minimal equipment, largely by hand. The terraces are located approximately 1500 metres (5000 ft) above sea level. They are fed by an ancient irrigation system from the rainforests above the terraces. It is said that if the steps were put end to end, it would encircle half the globe. The terraces are found in the province of Ifugao and the Ifugao people have been its caretakers. Ifugao culture revolves around rice and the culture displays an elaborate array of celebrations linked with agricultural rites from rice cultivation to rice consumption. The harvest season generally calls for thanksgiving feasts, while the concluding harvest rites called *tango* or *tungul* (a day of rest) entails a strict taboo on any agricultural work. Partaking of the *bayah* (rice beer), rice cakes, and betel nut constitutes an indelible practise during the festivities.

The banaue Rice Terraces in Ifugao, Philippines.

The Ifugao people practise traditional farming spending most of their labour at their terraces and forest lands while occasionally tending to root crop cultivation. The Ifugaos have also been known to culture edible shells, fruit trees, and other vegetables which have been exhibited among Ifugaos for generations. The building of the rice terraces consists of blanketing walls with stones and earth which are designed to draw water from a main irrigation canal above the terrace clusters. Indigenous rice terracing technologies have been identified with the Ifugao's rice terraces such as their knowledge of water irrigation, stonework, earthwork and terrace maintenance. As their source of life and art, the rice terraces have sustained and shaped the lives of the community members.

Sri Lanka

Rice is the staple food amongst all the ethnic groups in Sri Lanka. Agriculture in Sri Lanka mainly depends on the rice cultivation. Rice production is acutely dependent on rainfall and government supply necessity of water through irrigation channels throughout the cultivation seasons. The principal cultivation season, known as "Maha", is from October to March and the subsidiary cultivation season, known as "Yala", is from April to September. During Maha season, there is usually enough water to sustain the cultivation of all rice fields, nevertheless in Yala season there is only enough water for cultivation of half of the land extent.

Traditional rice varieties are now making a comeback with the recent interest in green foods.

Thailand

Rice is the main export of Thailand, especially white jasmine rice 105 (Dok Mali 105). Thailand has a large number of rice varieties, 3,500 kinds with different characters, and five kinds of wild rice cultivates. In each region of the country there are different rice seed types. Their use depends on weather, atmosphere, and topography.

The northern region has both low lands and high lands. The farmers' usual crop is non-glutinous rice such as Niew Sun Pah Tong rice. This rice is naturally protected from leaf disease, and its paddy (unmilled rice) (Thai: ข้าวเปลือก) has a brown color. The northeastern region is a large area where farmers can cultivate about 36 million square meters of rice. Although most of it is plains and dry areas, white jasmine rice 105—the most famous Thai rice—can be grown there. White jasmine rice was developed in Chonburi Province first and after that grown in many areas in the country, but the rice from this region has a high quality, because it's softer, whiter, and more fragrant. This rice can resist drought, acidic soil, and alkaline soil.

The central region is mostly composed of plains. Most farmers grow Jao rice. For example, Pathum Thani 1 rice which has qualities similar to white jasmine 105 rice. Its paddy has the color of thatch and the cooked rice has fragrant grains also.

In the southern region, most farmers transplant around boundaries to the flood plains or on the plains between mountains. Farming in the region is slower than other regions because the rainy season comes later. The popular rice varieties in this area are the Leb Nok Pattani seeds, a type of Jao rice. Its paddy has the color of thatch and it can be processed to make noodles.

Companion Plant

One of the earliest known examples of companion planting is the growing of rice with Azolla, the mosquito fern, which covers the top of a fresh rice paddy's water, blocking out any competing plants, as well as fixing nitrogen from the atmosphere for the rice to use. The rice is planted when it is tall enough to poke out above the azolla. This method has been used for at least a thousand years.

Middle East

Rice was grown in some areas of Mesopotamia (southern Iraq). With the rise of Islam it moved north to Nisibin, the southern shores of the Caspian Sea (in Gilan and Mazanderan provinces of Iran) and then beyond the Muslim world into the valley of the Volga. In Egypt, rice is mainly grown in the Nile Delta. In Palestine, rice came to be grown in the Jordan Valley. Rice is also grown in Saudi Arabia at Al-Hasa Oasis and in Yemen.

Europe

Rice was known to the Classical world, being imported from Egypt, and perhaps west Asia. It was known to Greece (where it is still cultivated in Macedonia and Thrace) by returning soldiers from Alexander the Great's military expedition to Asia. Large deposits of rice from the first century A.D. have been found in Roman camps in Germany.

The Moors brought Asiatic rice to the Iberian Peninsula in the 10th century. Records indicate it was grown in Valencia and Majorca. In Majorca, rice cultivation seems to have stopped after the Christian conquest, although historians are not certain.

Muslims also brought rice to Sicily, where it was an important crop long before it is noted in the plain of Pisa (1468) or in the Lombard plain (1475), where its cultivation was promoted by Ludovico Sforza, Duke of Milan, and demonstrated in his model farms.

After the 15th century, rice spread throughout Italy and then France, later propagating to all the continents during the age of European exploration.

In European Russia, a short-grain, starchy rice similar to the Italian varieties, has been grown in the Krasnodar Krai, and known in Russia as "Kuban Rice" or "Krasnodar Rice". In the Russian Far East several *japonica* cultivars are grown in Primorye around the Khanka lake. Increasing scale of rice production in the region has recently brought criticism towards growers' alleged bad practices in regards to the environment.

Caribbean and Latin America

Rice is not native to the Americas but was introduced to Latin America and the Caribbean by European colonizers at an early date. Spanish colonizers introduced Asian rice to Mexico in the 1520s at Veracruz; and the Portuguese and their African slaves introduced it at about the same time to colonial Brazil. Recent scholarship suggests that enslaved Africans played an active role in the establishment of rice in the New World and that African rice was an important crop from an early period. Varieties of rice and bean dishes that were a staple dish along the peoples of West Africa remained a staple among their descendants subjected to slavery in the Spanish New World colonies, Brazil and elsewhere in the Americas.

The Native Americans of what is now the Eastern United States may have practiced extensive agriculture with forms of wild rice (Zizania palustris), which looks similar to but it not directly related to rice.

United States

In 1694, rice arrived in South Carolina, probably originating from Madagascar.

In the United States, colonial South Carolina and Georgia grew and amassed great wealth from the slave labor obtained from the Senegambia area of West Africa and from coastal Sierra Leone. At the port of Charleston, through which 40% of all American slave imports passed, slaves from this region of Africa brought the highest prices due to their prior knowledge of rice culture, which was put to use on the many rice plantations around Georgetown, Charleston, and Savannah.

From the enslaved Africans, plantation owners learned how to dyke the marshes and periodically flood the fields. At first the rice was laboriously milled by hand using large mortars and pestles made of wood, then winnowed in sweetgrass baskets (the making of which was another skill brought by slaves from Africa). The invention of the rice mill increased profitability of the crop, and the addition of water power for the mills in 1787 by millwright Jonathan Lucas was another step forward.

Rice culture in the southeastern U.S. became less profitable with the loss of slave labor after the American Civil War, and it finally died out just after the turn of the 20th century. Today, people can visit the only remaining rice plantation in South Carolina that still has the original winnowing barn and rice mill from the mid-19th century at the historic Mansfield Plantation in Georgetown, South Carolina. The predominant strain of rice in the Carolinas was from Africa and was known as 'Carolina Gold'. The cultivar has been preserved and there are current attempts to reintroduce it as a commercially grown crop.

South Carolina rice plantation, showing a winnowing barn (Mansfield Plantation, Georgetown)

In the southern United States, rice has been grown in southern Arkansas, Louisiana, and east Texas since the mid-19th century. Many Cajun farmers grew rice in wet marshes and low-lying prairies where they could also farm crayfish when the fields were flooded. In recent years rice production has risen in North America, especially in the Mississippi embayment in the states of Arkansas and Mississippi.

Rice paddy fields just north of the city of Sacramento, California.

Rice cultivation began in California during the California Gold Rush, when an estimated 40,000 Chinese laborers immigrated to the state and grew small amounts of the grain for their own consumption. However, commercial production began only in 1912 in the town of Richvale in Butte County. By 2006, California produced the second-largest rice crop in the United States, after Arkansas, with production concentrated in six counties north of Sacramento. Unlike the Arkansas–Mississippi Delta region, California's production is dominated by short- and medium-grain *japonica* varieties, including cultivars developed for the local climate such as Calrose, which makes up as much as 85% of the state's crop.

References to wild rice in the Americas are to the unrelated *Zizania palustris*.

More than 100 varieties of rice are commercially produced primarily in six states (Arkansas, Texas, Louisiana, Mississippi, Missouri, and California) in the U.S. According to estimates for the 2006 crop year, rice production in the U.S. is valued at $1.88 billion, approximately half of which is expected to be exported. The U.S. provides about 12% of world rice trade. The majority of domestic utilization of U.S. rice is direct food use (58%), while 16% is used in each of processed foods and beer. 10% is found in pet food.

Australia

Rice was one of the earliest crops planted in Australia by British settlers, who had experience with rice plantations in the Americas and India.

Although attempts to grow rice in the well-watered north of Australia have been made for many years, they have consistently failed because of inherent iron and manganese toxicities in the soils and destruction by pests.

In the 1920s, it was seen as a possible irrigation crop on soils within the Murray-Darling Basin that were too heavy for the cultivation of fruit and too infertile for wheat.

Because irrigation water, despite the extremely low runoff of temperate Australia, was (and remains) very cheap, the growing of rice was taken up by agricultural groups over the following decades. Californian varieties of rice were found suitable for the climate in the Riverina, and the first mill opened at Leeton in 1951.

Monthly value (A$ millions) of rice imports to Australia since 1988

Even before this Australia's rice production greatly exceeded local needs, and rice exports to Japan have become a major source of foreign currency. Above-average rainfall from the 1950s to the middle 1990s encouraged the expansion of the Riverina rice industry, but its prodigious water use in a practically waterless region began to attract the attention of environmental scientists. These became severely concerned with declining flow in the Snowy River and the lower Murray River.

Although rice growing in Australia is highly profitable due to the cheapness of land, several recent years of severe drought have led many to call for its elimination because of its effects on extremely fragile aquatic ecosystems. The Australian rice industry is somewhat opportunistic, with the area planted varying significantly from season to season depending on water allocations in the Murray and Murrumbidgee irrigation regions.

Production and Commerce

Top 20 Rice Producers by Country—2012 (million metric ton)	
China	204.3
India	152.6
Indonesia	69.0

★	Vietnam	43.7
▬	Thailand	37.8
◯	Bangladesh	33.9
★	Myanmar	33.0
▶	Philippines	18.0
●	Brazil	11.5
●	Japan	10.7
☪	Pakistan	9.4
▦	Cambodia	9.3
▤	United States	9.0
☯	South Korea	6.4
▬	Egypt	5.9
▨	Nepal	5.1
▮ ▮	Nigeria	4.8
▬	Madagascar	4.0
▦	Sri Lanka	3.8
●	Laos	3.5

Production

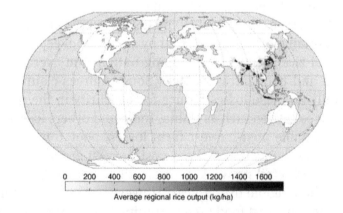

Worldwide rice production

The world dedicated 162.3 million hectares in 2012 for rice cultivation and the total production was about 738.1 million tonnes. The average world farm yield for rice was 4.5 tonnes per hectare, in 2012.

Rice farms in Egypt were the most productive in 2012, with a nationwide average of 9.5 tonnes per hectare. Second place: Australia – 8.9 tonnes per hectare. Third place: United States – 8.3 tonnes per hectare.

Burning of rice residues after harvest, to quickly prepare the land for wheat planting,
around Sangrur, Punjab, India.

Rice is a major food staple and a mainstay for the rural population and their food security. It is mainly cultivated by small farmers in holdings of less than 1 hectare. Rice is also a wage commodity for workers in the cash crop or non-agricultural sectors. Rice is vital for the nutrition of much of the population in Asia, as well as in Latin America and the Caribbean and in Africa; it is central to the food security of over half the world population. Developing countries account for 95% of the total production, with China and India alone responsible for nearly half of the world output.

World production of rice has risen steadily from about 200 million tonnes of paddy rice in 1960 to over 678 million tonnes in 2009. The three largest producers of rice in 2009 were China (197 million tonnes), India (131 Mt), and Indonesia (64 Mt). Among the six largest rice producers, the most productive farms for rice, in 2009, were in China producing 6.59 tonnes per hectare.

Many rice grain producing countries have significant losses post-harvest at the farm and because of poor roads, inadequate storage technologies, inefficient supply chains and farmer's inability to bring the produce into retail markets dominated by small shopkeepers. A World Bank – FAO study claims 8% to 26% of rice is lost in developing nations, on average, every year, because of post-harvest problems and poor infrastructure. Some sources claim the post-harvest losses to exceed 40%. Not only do these losses reduce food security in the world, the study claims that farmers in developing countries such as China, India and others lose approximately US$89 billion of income in preventable post-harvest farm losses, poor transport, the lack of proper storage and retail. One study claims that if these post-harvest grain losses could be eliminated with better infrastructure and retail network, in India alone enough food would be saved every year to feed 70 to 100 million people over a year. However, other writers have warned against dramatic assessments of post-harvest food losses, arguing that "worst-case scenarios" tend to be used rather than realistic averages and that in many cases the cost of avoiding losses exceeds the value of the food saved.

The seeds of the rice plant are first milled using a rice huller to remove the chaff (the outer husks of the grain). At this point in the process, the product is called brown rice. The milling may be continued, removing the bran, *i.e.*, the rest of the husk and the germ, thereby creating white rice. White rice, which keeps longer, lacks some important nutrients; moreover, in a limited diet which does not supplement the rice, brown rice helps to prevent the disease beriberi.

Either by hand or in a rice polisher, white rice may be buffed with glucose or talc powder (often called polished rice, though this term may also refer to white rice in general), parboiled, or processed into flour. White rice may also be enriched by adding nutrients, especially those lost during the milling process. While the cheapest method of enriching involves adding a powdered blend of nutrients that will easily wash off (in the United States, rice which has been so treated requires a label warning against rinsing), more sophisticated methods apply nutrients directly to the grain, coating the grain with a water-insoluble substance which is resistant to washing.

In some countries, a popular form, parboiled rice (also known as converted rice) is subjected to a steaming or parboiling process while still a brown rice grain. The parboil process causes a gelatinisation of the starch in the grains. The grains become less brittle, and the color of the milled grain changes from white to yellow. The rice is then dried, and can then be milled as usual or used as brown rice. Milled parboiled rice is nutritionally superior to standard milled rice, because the process causes nutrients from the outer husk (especially thiamine) to move into the endosperm, so that less is subsequently lost when the husk is polished off during milling. Parboiled rice has an additional benefit in that it does not stick to the pan during cooking, as happens when cooking regular white rice. This type of rice is eaten in parts of India and countries of West Africa are also accustomed to consuming parboiled rice.

Despite the hypothetical health risks of talc (such as stomach cancer), talc-coated rice remains the norm in some countries due to its attractive shiny appearance, but it has been banned in some, and is no longer widely used in others (such as the United States). Even where talc is not used, glucose, starch, or other coatings may be used to improve the appearance of the grains.

Rice bran, called *nuka* in Japan, is a valuable commodity in Asia and is used for many daily needs. It is a moist, oily inner layer which is heated to produce oil. It is also used as a pickling bed in making rice bran pickles and *takuan*.

Raw rice may be ground into flour for many uses, including making many kinds of beverages, such as *amazake, horchata*, rice milk, and rice wine. Rice does not contain gluten, so is suitable for people on a gluten-free diet. Rice may also be made into various types of noodles. Raw, wild, or brown rice may also be consumed by raw-foodist or fruitarians if soaked and sprouted (usually a week to 30 days – gaba rice).

Processed rice seeds must be boiled or steamed before eating. Boiled rice may be further fried in cooking oil or butter (known as fried rice), or beaten in a tub to make *mochi*.

Rice is a good source of protein and a staple food in many parts of the world, but it is not a complete protein: it does not contain all of the essential amino acids in sufficient amounts for good health, and should be combined with other sources of protein, such as nuts, seeds, beans, fish, or meat.

Rice, like other cereal grains, can be puffed (or popped). This process takes advantage of the grains' water content and typically involves heating grains in a special chamber. Further puffing is sometimes accomplished by processing puffed pellets in a low-pressure chamber. The ideal gas law means either lowering the local pressure or raising the water temperature results in an increase in volume prior to water evaporation, resulting in a puffy texture. Bulk raw rice density is about 0.9 g/cm^3. It decreases to less than one-tenth that when puffed.

Harvesting, Drying and Milling

Rice combine harvester Katori-city, Japan

Unmilled rice, known as "paddy" (Indonesia and Malaysia: padi; Philippines, palay), is usually harvested when the grains have a moisture content of around 25%. In most Asian countries, where rice is almost entirely the product of smallholder agriculture, harvesting is carried out manually, although there is a growing interest in mechanical harvesting. Harvesting can be carried out by the farmers themselves, but is also frequently done by seasonal labor groups. Harvesting is followed by threshing, either immediately or within a day or two. Again, much threshing is still carried out by hand but there is an increasing use of mechanical threshers. Subsequently, paddy needs to be dried to bring down the moisture content to no more than 20% for milling.

After the harvest, rice straw is gathered in the traditional way from small paddy
fields in Mae Wang District, Chiang Mai Province, Thailand

A familiar sight in several Asian countries is paddy laid out to dry along roads. However, in most countries the bulk of drying of marketed paddy takes place in mills, with village-level drying being used for paddy to be consumed by farm families. Mills either sun dry or use mechanical driers or both. Drying has to be carried out quickly to avoid the formation of molds. Mills range from simple hullers, with a throughput of a couple of tonnes a day, that simply remove the outer husk, to enormous operations that can process 4,000 tonnes a day and produce highly polished rice. A good mill can achieve a paddy-to-rice conversion rate of up to 72% but smaller, inefficient mills often struggle to achieve 60%. These smaller mills often do not buy paddy and sell rice but only service farmers who want to mill their paddy for their own consumption.

Distribution

Because of the importance of rice to human nutrition and food security in Asia, the domestic rice markets tend to be subject to considerable state involvement. While the private sector plays a leading role in most countries, agencies such as BULOG in Indonesia, the NFA in the Philippines, VINAFOOD in Vietnam and the Food Corporation of India are all heavily involved in purchasing of paddy from farmers or rice from mills and in distributing rice to poorer people. BULOG and NFA monopolise rice imports into their countries while VINAFOOD controls all exports from Vietnam.

Drying rice in Peravoor, India

Trade

World trade figures are very different from those for production, as less than 8% of rice produced is traded internationally. In economic terms, the global rice trade was a small fraction of 1% of world mercantile trade. Many countries consider rice as a strategic food staple, and various governments subject its trade to a wide range of controls and interventions.

Developing countries are the main players in the world rice trade, accounting for 83% of exports and 85% of imports. While there are numerous importers of rice, the exporters of rice are limited. Just five countries – Thailand, Vietnam, China, the United States and India – in decreasing order of exported quantities, accounted for about three-quarters of world rice exports in 2002. However, this ranking has been rapidly changing in recent years. In 2010, the three largest exporters of rice, in decreasing order of quantity exported were Thailand, Vietnam and India. By 2012, India became the largest exporter of rice with a 100% increase in its exports on year-to-year basis, and Thailand slipped to third position. Together, Thailand, Vietnam and India accounted for nearly 70% of the world rice exports.

The primary variety exported by Thailand and Vietnam were Jasmine rice, while exports from India included aromatic Basmati variety. China, an exporter of rice in early 2000s, was a net importer of rice in 2010 and will become the largest net importer, surpassing Nigeria, in 2013. According to a USDA report, the world's largest exporters of rice in 2012 were India (9.75 million tonnes), Vietnam (7 million tonnes), Thailand (6.5 million tonnes), Pakistan (3.75 million tonnes) and the United States (3.5 million tonnes).

Major importers usually include Nigeria, Indonesia, Bangladesh, Saudi Arabia, Iran, Iraq, Malaysia, the Philippines, Brazil and some African and Persian Gulf countries. In common with other West African countries, Nigeria is actively promoting domestic production. However, its very heavy import duties (110%) open it to smuggling from neighboring countries. Parboiled rice is particularly popular in Nigeria. Although China and India are the two largest producers of rice in the world, both countries consume the majority of the rice produced domestically, leaving little to be traded internationally.

World's Most Productive Rice Farms and Farmers

The average world yield for rice was 4.3 tonnes per hectare, in 2010.

Australian rice farms were the most productive in 2010, with a nationwide average of 10.8 tonnes per hectare.

Yuan Longping of China National Hybrid Rice Research and Development Center, China, set a world record for rice yield in 2010 at 19 tonnes per hectare on a demonstration plot. In 2011, this record was surpassed by an Indian farmer, Sumant Kumar, with 22.4 tonnes per hectare in Bihar. Both these farmers claim to have employed newly developed rice breeds and System of Rice Intensification (SRI), a recent innovation in rice farming. SRI is claimed to have set new national records in rice yields, within the last 10 years, in many countries. The claimed Chinese and Indian yields have yet to be demonstrated on seven-hectare lots and to be reproducible over two consecutive years on the same farm.

Price

In late 2007 to May 2008, the price of grains rose greatly due to droughts in major producing countries (particularly Australia), increased use of grains for animal feed and US subsidies for bio-fuel production. Although there was no shortage of rice on world markets this general upward trend in grain prices led to panic buying by consumers, government rice export bans (in particular, by Vietnam and India) and inflated import orders by the Philippines marketing board, the National Food Authority. This caused significant rises in rice prices. In late April 2008, prices hit 24 US cents a pound, twice the price of seven months earlier. Over the period of 2007 to 2013, the Chinese government has substantially increased the price it pays domestic farmers for their rice, rising to US$500 per metric ton by 2013. The 2013 price of rice originating from other southeast Asian countries was a comparably low US$350 per metric ton.

On April 30, 2008, Thailand announced plans for the creation of the Organisation of Rice Exporting Countries (OREC) with the intention that this should develop into a price-fixing cartel for rice. However, little progress had been made by mid-2011 to achieve this.

Worldwide Consumption

Food consumption of rice by country – 2009 (million metric ton of paddy equivalent)	
World Total	531.6
China	156.3
India	123.5

	Bangladesh	50.4
	Indonesia	45.3
	Vietnam	18.4
	Philippines	17.0
	Thailand	13.7
	Japan	10.2
	Burma	10.0
	Brazil	10.0
	South Korea	5.8
	Nigeria	4.8
	Egypt	4.6
	Pakistan	4.3
	USA	3.8
	Nepal	3.5
	Cambodia	3.4
	Sri Lanka	3.2
	Madagascar	3.2
	Malaysia	3.1
	North Korea	2.8

As of 2009 world food consumption of rice was 531.6 million metric tons of paddy equivalent (354,603 of milled equivalent), while the far largest consumers were China consuming 156.3 million metric tons of paddy equivalent (29.4% of the world consumption) and India consuming 123.5 million metric tons of paddy equivalent (23.3% of the world consumption). Between 1961 and 2002, per capita consumption of rice increased by 40%.

Rice is the most important crop in Asia. In Cambodia, for example, 90% of the total agricultural area is used for rice production.

U.S. rice consumption has risen sharply over the past 25 years, fueled in part by commercial applications such as beer production. Almost one in five adult Americans now report eating at least half a serving of white or brown rice per day.

Environmental Impacts

Rice cultivation on wetland rice fields is thought to be responsible for 11% of the anthropogenic methane emissions. Rice requires slightly more water to produce than other grains. Rice production uses almost a third of Earth's fresh water.

Long-term flooding of rice fields cuts the soil off from atmospheric oxygen and causes anaerobic fermentation of organic matter in the soil. Methane production from rice cultivation contributes ~1.5% of anthropogenic greenhouse gases. Methane is twenty times more potent a greenhouse gas than carbon dioxide.

Work by the International Center for Tropical Agriculture to measure the greenhouse gas emissions of rice production.

A 2010 study found that, as a result of rising temperatures and decreasing solar radiation during the later years of the 20th century, the rice yield growth rate has decreased in many parts of Asia, compared to what would have been observed had the temperature and solar radiation trends not occurred. The yield growth rate had fallen 10–20% at some locations. The study was based on records from 227 farms in Thailand, Vietnam, Nepal, India, China, Bangladesh, and Pakistan. The mechanism of this falling yield was not clear, but might involve increased respiration during warm nights, which expends energy without being able to photosynthesize.

Rainfall

Temperature

Rice requires high temperature above 20 °C but not more than 35 to 40 °C. Optimum temperature is around 30 °C (T_{max}) and 20 °C (T_{min}).

Solar Radiation

The amount of solar radiation received during 45 days after harvest determines final crop output.

Atmospheric Water Vapor

High water vapor content (in humid tropics) subjects unusual stress which favors the spread of fungal and bacterial diseases.

Wind

Light wind transports CO_2 to the leaf canopy but strong wind cause severe damage and may lead to sterility (due to pollen dehydration, spikelet sterility, and abortive endosperms).

Pests and Diseases

Rice pests are any organisms or microbes with the potential to reduce the yield or value of the rice crop (or of rice seeds). Rice pests include weeds, pathogens, insects, nematode, rodents, and birds. A variety of factors can contribute to pest outbreaks, including climatic factors, improper irrigation, the overuse of insecticides and high rates of nitrogen fertilizer application. Weather conditions also contribute to pest outbreaks. For example, rice gall midge and army worm outbreaks tend to follow periods of high rainfall early in the wet season, while thrips outbreaks are associated with drought.

Insects

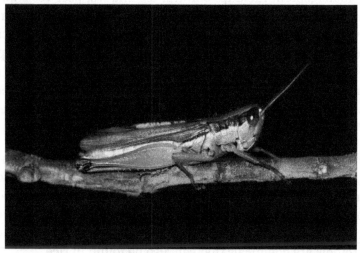

Chinese rice grasshopper (*Oxya chinensis*) Borneo, Malaysia

Major rice insect pests include: the brown planthopper (BPH), several spp. of stemborers – including those in the genera *Scirpophaga* and *Chilo*, the rice gall midge, several spp. of rice bugs – notably in the genus *Leptocorisa*, the rice leafroller, rice weevils and the Chinese rice grasshopper.

Diseases

Rice blast, caused by the fungus *Magnaporthe grisea*, is the most significant disease affecting rice cultivation. Other major rice diseases include: sheath blight, rice ragged stunt (vector: BPH), and tungro (vector: *Nephotettix* spp). There is also an ascomycete fungus, *Cochliobolus miyabeanus*, that causes brown spot disease in rice.

Nematodes

Several nematode species infect rice crops, causing diseases such as Ufra (Ditylenchus dipsaci), White tip disease (Aphelenchoide bessei), and root knot disease (Meloidogyne graminicola). Some nematode species such as *Pratylenchus* spp. are most dangerous in upland rice of all parts of the world. Rice root nematode (*Hirschmanniella oryzae*) is a migratory endoparasite which on higher inoculum levels will lead to complete destruction of a rice crop. Beyond being obligate parasites, they also decrease the vigor of plants and increase the plants' susceptibility to other pests and diseases.

Other Pests

These include the apple snail *Pomacea canaliculata*, panicle rice mite, rats, and the weed *Echinochloa crusgali*.

Integrated Pest Management

Crop protection scientists are trying to develop rice pest management techniques which are sustainable. In other words, to manage crop pests in such a manner that future crop production is not threatened. Sustainable pest management is based on four principles: biodiversity, host plant resistance (HPR), landscape ecology, and hierarchies in a landscape – from biological to social. At present, rice pest management includes cultural techniques, pest-resistant rice varieties, and pesticides (which include insecticide). Increasingly, there is evidence that farmers' pesticide applications are often unnecessary, and even facilitate pest outbreaks. By reducing the populations of natural enemies of rice pests, misuse of insecticides can actually lead to pest outbreaks. The International Rice Research Institute (IRRI) demonstrated in 1993 that an 87.5% reduction in pesticide use can lead to an overall drop in pest numbers. IRRI also conducted two campaigns in 1994 and 2003, respectively, which discouraged insecticide misuse and smarter pest management in Vietnam.

Rice plants produce their own chemical defenses to protect themselves from pest attacks. Some synthetic chemicals, such as the herbicide 2,4-D, cause the plant to increase the production of certain defensive chemicals and thereby increase the plant's resistance to some types of pests. Conversely, other chemicals, such as the insecticide imidacloprid, can induce changes in the gene expression of the rice that cause the plant to become more susceptible to attacks by certain types of pests. 5-Alkylresorcinols are chemicals that can also be found in rice.

Chloroxylon is used for Pest Management in Organic Rice Cultivation in Chhattisgarh, India

Botanicals, so-called "natural pesticides", are used by some farmers in an attempt to control rice pests. Botanicals include extracts of leaves, or a mulch of the leaves themselves. Some upland rice farmers in Cambodia spread chopped leaves of the bitter bush (*Chromolaena odorata*) over the surface of fields after planting. This practice probably helps the soil retain moisture and thereby facilitates seed germination. Farmers also claim the leaves are a natural fertilizer and helps suppress weed and insect infestations.

Among rice cultivars, there are differences in the responses to, and recovery from, pest damage. Many rice varieties have been selected for resistance to insect pests. Therefore, particular cultivars are recommended for areas prone to certain pest problems. The genetically based ability of a rice variety to withstand pest attacks is called resistance. Three main types of plant resistance to pests are recognized as nonpreference, antibiosis, and tolerance. Nonpreference (or antixenosis) describes host plants which insects prefer to avoid; antibiosis is where insect survival is reduced after the ingestion of host tissue; and tolerance is the capacity of a plant to produce high yield or retain high quality despite insect infestation.

Over time, the use of pest-resistant rice varieties selects for pests that are able to overcome these mechanisms of resistance. When a rice variety is no longer able to resist pest infestations, resistance is said to have broken down. Rice varieties that can be widely grown for many years in the presence of pests and retain their ability to withstand the pests are said to have durable resistance. Mutants of popular rice varieties are regularly screened by plant breeders to discover new sources of durable resistance.

Parasitic Weeds

Rice is parasitized by the weed eudicot *Striga hermonthica*, which is of local importance for this crop.

Ecotypes and Cultivars

While most rice is bred for crop quality and productivity, there are varieties selected for characteristics such as texture, smell, and firmness. There are four major categories of rice worldwide: indica, japonica, aromatic and glutinous. The different varieties of rice are not considered interchangeable, either in food preparation or agriculture, so as a result, each major variety is a completely separate market from other varieties. It is common for one variety of rice to rise in price while another one drops in price.

Rice cultivars also fall into groups according to environmental conditions, season of planting, and season of harvest, called ecotypes. Some major groups are the Japan-type (grown in Japan), "buly" and "tjereh" types (Indonesia); "aman" (main winter crop), "aus" ("aush", summer), and "boro" (spring) (Bengal and Assam). Cultivars exist that are adapted to deep flooding, and these are generally called "floating rice".

The largest collection of rice cultivars is at the International Rice Research Institute in the Philippines, with over 100,000 rice accessions held in the International Rice Genebank. Rice cultivars are often classified by their grain shapes and texture. For example, Thai Jasmine rice is long-grain and relatively less sticky, as some long-grain rice contains less amylopectin than short-grain cultivars. Chinese restaurants often serve long-grain as plain unseasoned steamed rice though short-grain rice is common as well. Japanese mochi rice and Chinese sticky rice are short-grain. Chinese people use sticky rice which is properly known as "glutinous rice" (note: glutinous refer to the glue-like characteristic of rice; does not refer to "gluten") to make zongzi. The Japanese table rice is a sticky, short-grain rice. Japanese sake rice is another kind as well.

Rice seed collection from IRRI

Indian rice cultivars include long-grained and aromatic Basmati (ਬਾਸਮਤੀ) (grown in the North), long and medium-grained Patna rice, and in South India (Andhra Pradesh and Karnataka) short-grained Sona Masuri (also called as Bangaru theegalu). In the state of Tamil Nadu, the most prized cultivar is *ponni* which is primarily grown in the delta regions of the Kaveri River. Kaveri is also referred to as ponni in the South and the name reflects the geographic region where it is grown. In the Western Indian state of Maharashtra, a short grain variety called Ambemohar is very popular. This rice has a characteristic fragrance of Mango blossom.

Aromatic rices have definite aromas and flavors; the most noted cultivars are Thai fragrant rice, Basmati, Patna rice, Vietnamese fragrant rice, and a hybrid cultivar from America, sold under the trade name Texmati. Both Basmati and Texmati have a mild popcorn-like aroma and flavor. In Indonesia, there are also *red* and *black* cultivars.

High-yield cultivars of rice suitable for cultivation in Africa and other dry ecosystems, called the new rice for Africa (NERICA) cultivars, have been developed. It is hoped that their cultivation will improve food security in West Africa.

Draft genomes for the two most common rice cultivars, *indica* and *japonica*, were published in April 2002. Rice was chosen as a model organism for the biology of grasses because of its relatively small genome (~430 megabase pairs). Rice was the first crop with a complete genome sequence.

On December 16, 2002, the UN General Assembly declared the year 2004 the International Year of Rice. The declaration was sponsored by more than 40 countries.

Biotechnology

High-yielding Varieties

The high-yielding varieties are a group of crops created intentionally during the Green Revolution to increase global food production. This project enabled labor markets in Asia to shift away from agriculture, and into industrial sectors. The first "Rice Car", IR8 was produced in 1966 at the International Rice Research Institute which is based in the Philippines at the University of the Philippines' Los Baños site. IR8 was created through a cross between an Indonesian variety named "Peta" and a Chinese variety named "Dee Geo Woo Gen."

Scientists have identified and cloned many genes involved in the gibberellin signaling pathway, including GAI1 (Gibberellin Insensitive) and SLR1 (Slender Rice). Disruption of gibberellin signaling can lead to significantly reduced stem growth leading to a dwarf phenotype. Photosynthetic investment in the stem is reduced dramatically as the shorter plants are inherently more stable mechanically. Assimilates become redirected to grain production, amplifying in particular the effect of chemical fertilizers on commercial yield. In the presence of nitrogen fertilizers, and intensive crop management, these varieties increase their yield two to three times.

Future Potential

As the UN Millennium Development project seeks to spread global economic development to Africa, the "Green Revolution" is cited as the model for economic development. With the intent of replicating the successful Asian boom in agronomic productivity, groups like the Earth Institute are doing research on African agricultural systems, hoping to increase productivity. An important way this can happen is the production of "New Rices for Africa" (NERICA). These rices, selected to tolerate the low input and harsh growing conditions of African agriculture, are produced by the African Rice Center, and billed as technology "from Africa, for Africa". The NERICA have appeared in *The New York Times* (October 10, 2007) and *International Herald Tribune* (October 9, 2007), trumpeted as miracle crops that will dramatically increase rice yield in Africa and enable an economic resurgence. Ongoing research in China to develop perennial rice could result in enhanced sustainability and food security.

Golden Rice

Rice kernels do not contain vitamin A, so people who obtain most of their calories from rice are at risk of vitamin A deficiency. German and Swiss researchers have genetically engineered rice to produce beta-carotene, the precursor to vitamin A, in the rice kernel. The beta-carotene turns the processed (white) rice a "gold" color, hence the name "golden rice." The beta-carotene is converted to vitamin A in humans who consume the rice. Although some rice strains produce beta-carotene in the hull, no non-genetically engineered strains have been found that produce beta-carotene in the kernel, despite the testing of thousands of strains. Additional efforts are being made to improve the quantity and quality of other nutrients in golden rice.

The International Rice Research Institute is currently further developing and evaluating Golden Rice as a potential new way to help address vitamin A deficiency.

Expression of Human Proteins

Ventria Bioscience has genetically modified rice to express lactoferrin, lysozyme which are proteins usually found in breast milk, and human serum albumin, These proteins have antiviral, antibacterial, and antifungal effects.

Rice containing these added proteins can be used as a component in oral rehydration solutions which are used to treat diarrheal diseases, thereby shortening their duration and reducing recurrence. Such supplements may also help reverse anemia.

Flood-tolerant Rice

Due to the varying levels that water can reach in regions of cultivation, flood tolerant varieties have long been developed and used. Flooding is an issue that many rice growers face, especially in South and South East Asia where flooding annually affects 20 million hectares. Standard rice varieties cannot withstand stagnant flooding of more than about a week, mainly as it disallows the plant access to necessary requirements such as sunlight and essential gas exchanges, inevitably leading to plants being unable to recover. In the past, this has led to massive losses in yields, such as in the Philippines, where in 2006, rice crops worth $65 million were lost to flooding. Recently developed cultivars seek to improve flood tolerance.

Drought-tolerant Rice

Drought represents a significant environmental stress for rice production, with 19–23 million hectares of rainfed rice production in South and South East Asia often at risk. Under drought conditions, without sufficient water to afford them the ability to obtain the required levels of nutrients from the soil, conventional commercial rice varieties can be severely affected – for example, yield losses as high as 40% have affected some parts of India, with resulting losses of around US$800 million annually.

The International Rice Research Institute conducts research into developing drought-tolerant rice varieties, including the varieties 5411 and Sookha dhan, currently being employed by farmers in the Philippines and Nepal respectively. In addition, in 2013 the Japanese National Institute for Agrobiological Sciences led a team which successfully inserted the DEEPER ROOTING 1 (DRO1) gene, from the Philippine upland rice variety Kinandang Patong, into the popular commercial rice variety IR64, giving rise to a far deeper root system in the resulting plants. This facilitates an improved ability for the rice plant to derive its required nutrients in times of drought via accessing deeper layers of soil, a feature demonstrated by trials which saw the IR64 + DRO1 rice yields drop by 10% under moderate drought conditions, compared to 60% for the unmodified IR64 variety.

Salt-tolerant Rice

Soil salinity poses a major threat to rice crop productivity, particularly along low-lying coastal areas during the dry season. For example, roughly 1 million hectares of the coastal areas of Bangladesh are affected by saline soils. These high concentrations of salt can severely affect rice plants' normal physiology, especially during early stages of growth, and as such farmers are often forced to abandon these otherwise potentially usable areas.

Progress has been made, however, in developing rice varieties capable of tolerating such conditions; the hybrid created from the cross between the commercial rice variety IR56 and the wild rice species *Oryza coarctata* is one example. *O. coarctata* is capable of successful growth in soils with double the limit of salinity of normal varieties, but lacks the ability to produce edible rice. Developed by the International Rice Research Institute, the hybrid variety can utilise specialised leaf glands that allow for the removal of salt into the atmosphere. It was initially produced from one successful embryo out of 34,000 crosses between the two species; this was then backcrossed to IR56 with the aim of preserving the genes responsible for salt tolerance that were inherited from *O. coarctata*. Extensive trials are planned prior to the new variety being available to farmers by approximately 2017–18.

Environment-friendly Rice

Producing rice in paddies is harmful for the environment due to the release of methane by methanogenic bacteria. These bacteria live in the anaerobic waterlogged soil, and live off nutrients released by rice roots. Researchers have recently reported in *Nature* that putting the barley gene SUSIBA2 into rice creates a shift in biomass production from root to shoot (above ground tissue becomes larger, while below ground tissue is reduced), decreasing the methanogen population, and resulting in a reduction of methane emissions of up to 97%. Apart from this environmental benefit, the modification also increases the amount of rice grains by 43%, which makes it useful tool in feeding a growing world population.

Meiosis and DNA Repair

Rice is used as a model organism for investigating the molecular mechanisms of meiosis and DNA repair in higher plants. Meiosis is a key stage of the sexual cycle in which diploid cells in the ovule (female structure) and the anther (male structure) produce haploid cells that develop further into gametophytes and gametes. So far, 28 meiotic genes of rice have been characterized. Studies of rice gene OsRAD51C showed that this gene is necessary for homologous recombinational repair of DNA, particularly the accurate repair of DNA double-strand breaks during meiosis. Rice gene OsDMC1 was found to be essential for pairing of homologous chromosomes during meiosis, and rice gene OsMRE11 was found to be required for both synapsis of homologous chromosomes and repair of double-strand breaks during meiosis.

Cultural Roles of Rice

Ancient statue of Dewi Sri from Java (c. 9th century)

Rice plays an important role in certain religions and popular beliefs. In many cultures relatives will scatter rice during or towards the end of a wedding ceremony in front of the bride and groom.

The pounded rice ritual is conducted during weddings in Nepal. The bride gives a leafplate full of pounded rice to the groom after he requests it politely from her.

In the Philippines rice wine, popularly known as *tapuy*, is used for important occasions such as weddings, rice harvesting ceremonies and other celebrations.

Dewi Sri is the traditional rice goddess of the Javanese, Sundanese, and Balinese people in Indonesia. Most rituals involving Dewi Sri are associated with the mythical origin attributed to the rice plant, the staple food of the region. In Thailand a similar rice deity is known as *Phosop*; she is a deity more related to ancient local folklore than a goddess of a structured, mainstream religion. The same female rice deity is known as *Po Ino Nogar* in Cambodia and as *Nang Khosop* in Laos. Ritual offerings are made during the different stages of rice production to propitiate the Rice Goddess in the corresponding cultures.

Maize

Maize (**MAYZ**; *Zea mays* subsp. *mays*, from Spanish: *maíz* after Taíno *mahiz*), also known as corn, is a large grain plant first domesticated by indigenous peoples in Mexico about 10,000 years ago. The six major types of corn are dent corn, flint corn, pod corn, popcorn, flour corn, and sweet corn.

The leafy stalk of the plant produces separate pollen and ovuliferous inflorescences or ears, which are fruits, yielding kernels (often erroneously called seeds). Maize kernels are often used in cooking as a starch.

History

Guilá Naquitz Cave in Oaxaca, Mexico is the site of early domestication of several food crops, including teosinte (an ancestor of maize).

Most historians believe maize was domesticated in the Tehuacan Valley of Mexico. Recent research modified this view somewhat; scholars now indicate the adjacent Balsas River Valley of south-central Mexico as the center of domestication.

The Olmec and Mayans cultivated maize in numerous varieties throughout Mesoamerica, cooked, ground or processed through nixtamalization. Beginning about 2500 BC, the crop spread through much of the Americas. The region developed a trade network based on surplus and varieties of maize crops.

Nevertheless, recent data indicates that the spread of maize took place even earlier. According to Piperno,

"A large corpus of data indicates that it [maize] was dispersed into lower Central America by 7600 BP [5600 BC] and had moved into the inter-Andean valleys of Colombia between 7000 and 6000 BP [5000-4000 BC]."

Since then, even earlier dates have been published.

After European contact with the Americas in the late 15th and early 16th centuries, explorers and traders carried maize back to Europe and introduced it to other countries. Maize spread to the rest of the world because of its ability to grow in diverse climates. Sugar-rich varieties called sweet corn are usually grown for human consumption as kernels, while field corn varieties are used for animal feed, various corn-based human food uses (including grinding into cornmeal or masa, pressing into corn oil, and fermentation and distillation into alcoholic beverages like bourbon whiskey), and as chemical feedstocks.

An influential 2002 study by Matsuoka *et al.* has demonstrated that, rather than the multiple independent domestications model, all maize arose from a single domestication in southern Mexico about 9,000 years ago. The study also demonstrated that the oldest surviving maize types are those of the Mexican highlands. Later, maize spread from this region over the Americas along two major paths. This is consistent with a model based on the archaeological record suggesting that maize diversified in the highlands of Mexico before spreading to the lowlands.

Before they were domesticated, maize plants only grew small, 25 millimetres (1 in) long corn cobs, and only one per plant. Many centuries of artificial selection by the indigenous people of the Americas resulted in the development of maize plants capable of growing several cobs per plant that were usually several centimetres/inches long each.

Maize is the most widely grown grain crop throughout the Americas, with 332 million metric tons grown annually in the United States alone. Approximately 40% of the crop—130 million tons—is used for corn ethanol. Genetically modified maize made up 85% of the maize planted in the United States in 2009.

Names

The word *maize* derives from the Spanish form of the indigenous Taíno word for the plant, *mahiz*. It is known by other names around the world.

The word "corn" outside North America, Australia, and New Zealand refers to any cereal crop, its meaning understood to vary geographically to refer to the local staple. In the United States, Canada,

Australia, and New Zealand, *corn* primarily means maize; this usage started as a shortening of "Indian corn". "Indian corn" primarily means maize (the staple grain of indigenous Americans), but can refer more specifically to multicolored "flint corn" used for decoration.

Many small male flowers make up the male inflorescence, called the tassel.

In places outside North America, Australia, and New Zealand, *corn* often refers to maize in culinary contexts. The narrower meaning is usually indicated by some additional word, as in *sweet corn*, *sweetcorn*, *corn on the cob*, *baby corn*, the puffed confection known as *popcorn* and the breakfast cereal known as *corn flakes*.

In Southern Africa, maize is commonly called *mielie* (Afrikaans) or *mealie* (English), words derived from the Portuguese word for maize, *milho*.

Maize is preferred in formal, scientific, and international usage because it refers specifically to this one grain, unlike *corn*, which has a complex variety of meanings that vary by context and geographic region. *Maize* is used by agricultural bodies and research institutes such as the FAO and CSIRO. National agricultural and industry associations often include the word *maize* in their name even in English-speaking countries where the local, informal word is something other than *maize*; for example, the Maize Association of Australia, the Indian Maize Development Association, the Kenya Maize Consortium and Maize Breeders Network, the National Maize Association of Nigeria, the Zimbabwe Seed Maize Association. However, in commodities trading, *corn* consistently refers to maize and not other grains.

Structure and Physiology

The maize plant is often 3 m (10 ft) in height, though some natural strains can grow 12 m (39 ft). The stem is commonly composed of 20 internodes of 18 cm (7.1 in) length. A leaf, which grows from each node, is generally 9 cm (4 in) in width and 120 cm (4 ft) in length.

Ears develop above a few of the leaves in the midsection of the plant, between the stem and leaf sheath, elongating by ~3 mm/day, to a length of 18 cm (7 in) with 60 cm (24 in) being the maximum alleged in the subspecies. They are female inflorescences, tightly enveloped by several layers of ear leaves commonly called husks. Certain varieties of maize have been bred to produce many additional developed ears. These are the source of the "baby corn" used as a vegetable in Asian cuisine.

The apex of the stem ends in the tassel, an inflorescence of male flowers. When the tassel is mature and conditions are suitably warm and dry, anthers on the tassel dehisce and release pollen. Maize pollen is anemophilous (dispersed by wind), and because of its large settling velocity, most pollen falls within a few meters of the tassel.

Female inflorescence, with young silk

Elongated stigmas, called silks, emerge from the whorl of husk leaves at the end of the ear. They are often pale yellow and 18 cm (7 in) in length, like tufts of hair in appearance. At the end of each is a carpel, which may develop into a "kernel" if fertilized by a pollen grain. The pericarp of the fruit is fused with the seed coat referred to as "caryopsis", typical of the grasses, and the entire kernel is often referred to as the "seed". The cob is close to a multiple fruit in structure, except that the individual fruits (the kernels) never fuse into a single mass. The grains are about the size of peas, and adhere in regular rows around a white, pithy substance, which forms the ear- maximum size of kernel in subspecies is reputedly 2.5 cm (1 in). An ear commonly holds 600 kernels. They are of various colors: blackish, bluish-gray, purple, green, red, white and yellow. When ground into flour, maize yields more flour with much less bran than wheat does. It lacks the protein gluten of wheat and, therefore, makes baked goods with poor rising capability. A genetic variant that accumulates more sugar and less starch in the ear is consumed as a vegetable and is called sweet corn. Young ears can be consumed raw, with the cob and silk, but as the plant matures (usually during the summer months), the cob becomes tougher and the silk dries to inedibility. By the end of the growing season, the kernels dry out and become difficult to chew without cooking them tender first in boiling water.

mature silk

Planting density affects multiple aspects of maize. Modern farming techniques in developed countries usually rely on dense planting, which produces one ear per stalk. Stands of silage maize are yet denser, and achieve a lower percentage of ears and more plant matter.

Stalks, ears, and silk

Male flowers

Maize is a facultative short-day plant and flowers in a certain number of growing degree days > 10 °C (50 °F) in the environment to which it is adapted. The magnitude of the influence that long nights have on the number of days that must pass before maize flowers is genetically prescribed and regulated by the phytochrome system. Photoperiodicity can be eccentric in tropical cultivars such that the long days characteristic of higher latitudes allow the plants to grow so tall that they do not have enough time to produce seed before being killed by frost. These attributes, however, may prove useful in using tropical maize for biofuels.

Full-grown maize plants

Mature maize ear on a stalk

Immature maize shoots accumulate a powerful antibiotic substance, 2,4-dihydroxy-7-methoxy-1,4-benzoxazin-3-one (DIMBOA). DIMBOA is a member of a group of hydroxamic acids (also known as benzoxazinoids) that serve as a natural defense against a wide range of pests, including insects, pathogenic fungi and bacteria. DIMBOA is also found in related grasses, particularly wheat. A maize mutant (bx) lacking DIMBOA is highly susceptible to attack by aphids and fungi. DIMBOA is also responsible for the relative resistance of immature maize to the European corn borer (family Crambidae). As maize matures, DIMBOA levels and resistance to the corn borer decline.

Because of its shallow roots, maize is susceptible to droughts, intolerant of nutrient-deficient soils, and prone to be uprooted by severe winds.

Maize kernels

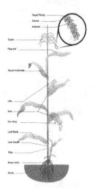

Maize plant diagram

Ear of maize with irregular rows of seeds

Zea mays "strawberry"—MHNT

Zea mays "Oaxacan Green" MHNT

Variegated maize ears

Multicolored corn kernels (CSIRO)

While yellow maizes derive their color from lutein and zeaxanthin, in red-colored maizes, the kernel coloration is due to anthocyanins and phlobaphenes. These latter substances are synthesized in the flavonoids synthetic pathway from polymerisation of flavan-4-ols by the expression of maize pericarp color1 (p1) gene which encodes an R2R3 myb-like transcriptional activator of the A1 gene encoding for the dihydroflavonol 4-reductase (reducing dihydroflavonols into flavan-4-ols) while another gene (Suppressor of Pericarp Pigmentation 1 or SPP1) acts as a suppressor. The p1 gene encodes an Myb-homologous transcriptional activator of genes required for biosynthesis of red phlobaphene pigments, while the P1-wr allele specifies colorless kernel pericarp and red cobs, and unstable factor for orange1 (Ufo1) modifies P1-wr expression to confer pigmentation in kernel pericarp, as well as vegetative tissues, which normally do not accumulate significant amounts of phlobaphene pigments. The maize P gene encodes a Myb homolog that recognizes the sequence CCT/AACC, in sharp contrast with the C/TAACGG bound by vertebrate Myb proteins.

Abnormal Flowers

Sometimes in maize, inflorescences are found containing both male and female flowers, or hermaphrodite flowers.

Genetics

Exotic varieties of maize are collected to add genetic diversity when selectively breeding new domestic strains

Many forms of maize are used for food, sometimes classified as various subspecies related to the amount of starch each has:

- Flour corn: *Zea mays* var. *amylacea*
- Popcorn: *Zea mays* var. *everta*
- Dent corn : *Zea mays* var. *indentata*
- Flint corn: *Zea mays* var. *indurata*
- Sweet corn: *Zea mays* var. *saccharata* and *Zea mays* var. *rugosa*
- Waxy corn: *Zea mays* var. *ceratina*
- Amylomaize: *Zea mays*
- Pod corn: *Zea mays* var. *tunicata* Larrañaga ex A. St. Hil.
- Striped maize: *Zea mays* var. *japonica*

This system has been replaced (though not entirely displaced) over the last 60 years by multivariable classifications based on ever more data. Agronomic data were supplemented by botanical traits for a robust initial classification, then genetic, cytological, protein and DNA evidence was added. Now, the categories are forms (little used), races, racial complexes, and recently branches.

Maize is a diploid with 20 chromosomes (n=10). The combined length of the chromosomes is 1500 cM. Some of the maize chromosomes have what are known as "chromosomal knobs": highly repetitive heterochromatic domains that stain darkly. Individual knobs are polymorphic among strains of both maize and teosinte.

Barbara McClintock used these knob markers to validate her transposon theory of "jumping genes", for which she won the 1983 Nobel Prize in Physiology or Medicine. Maize is still an important model organism for genetics and developmental biology today.

The Maize Genetics Cooperation Stock Center, funded by the USDA Agricultural Research Service and located in the Department of Crop Sciences at the University of Illinois at Urbana-Champaign, is a stock center of maize mutants. The total collection has nearly 80,000 samples. The bulk of the collection consists of several hundred named genes, plus additional gene combinations and other heritable variants. There are about 1000 chromosomal aberrations (e.g., translocations and inversions) and stocks with abnormal chromosome numbers (e.g., tetraploids). Genetic data describing the maize mutant stocks as well as myriad other data about maize genetics can be accessed at MaizeGDB, the Maize Genetics and Genomics Database.

In 2005, the US National Science Foundation (NSF), Department of Agriculture (USDA) and the Department of Energy (DOE) formed a consortium to sequence the B73 maize genome. The resulting DNA sequence data was deposited immediately into GenBank, a public repository for genome-sequence data. Sequences and genome annotations have also been made available throughout the project's lifetime at the project's official site.

Primary sequencing of the maize genome was completed in 2008. On November 20, 2009, the consortium published results of its sequencing effort in *Science*. The genome, 85% of which is composed of transposons, was found to contain 32,540 genes (By comparison, the human genome contains about 2.9 billion bases and 26,000 genes). Much of the maize genome has been duplicated and reshuffled by helitrons—group of rolling circle transposons.

According to a genetic study by Embrapa, corn cultivation was introduced in South America from Mexico, in two great waves: the first, 5000 years ago, spread through the Andes; the second, about 2000 years ago, through the lowlands of South America.

Breeding

Maize reproduces sexually each year. This randomly selects half the genes from a given plant to propagate to the next generation, meaning that desirable traits found in the crop (like high yield or good nutrition) can be lost in subsequent generations unless certain techniques are used.

Maize breeding in prehistory resulted in large plants producing large ears. Modern breeding began with individuals who selected highly productive varieties in their fields and then sold seed to other farmers. James L. Reid was one of the earliest and most successful developing Reid's Yellow Dent in the 1860s. These early efforts were based on mass selection. Later breeding efforts included ear to row selection, (C. G. Hopkins ca. 1896), hybrids made from selected inbred lines (G. H. Shull, 1909), and the highly successful double cross hybrids using 4 inbred lines (D. F. Jones ca. 1918, 1922). University supported breeding programs were especially important in developing and introducing modern hybrids. (Ref Jugenheimer Hybrid Maize Breeding and Seed Production pub.

1958) by the 1930s, companies such as Pioneer devoted to production of hybrid maize had begun to influence long term development. Internationally important seed banks such as International Maize and Wheat Improvement Center (CIMMYT) and the US bank at Maize Genetics Cooperation Stock Center University of Illinois at Urbana-Champaign maintain germplasm important for future crop development.

Field of maize in Liechtenstein

Since the 1940s the best strains of maize have been first-generation hybrids made from inbred strains that have been optimized for specific traits, such as yield, nutrition, drought, pest and disease tolerance. Both conventional cross-breeding and genetic modification have succeeded in increasing output and reducing the need for cropland, pesticides, water and fertilizer.

Global Maize Program

Panorama of cornfields in Nan Province, Thailand

CIMMYT operates a conventional breeding program to provide optimized strains. The program began in the 1980s. Hybrid seeds are distributed in Africa by the Drought Tolerant Maize for Africa project.

Genetic Modification

Genetically modified (GM) maize is one of the 25 GM crops grown commercially in 2011. Grown since 1997 in the United States and Canada, 86% of the US maize crop was genetically modified in

2010 and 32% of the worldwide maize crop was GM in 2011. As of 2011, Herbicide-tolerant maize varieties are grown in Argentina, Australia, Brazil, Canada, China, Colombia, El Salvador, the EU, Honduras, Japan, Korea, Malaysia, Mexico, New Zealand, Philippines, the Russian Federation, Singapore, South Africa, Taiwan, Thailand, and USA, and insect-resistant corn is grown in Argentina, Australia, Brazil, Canada, Chile, China, Colombia, Czech Republic, Egypt, the EU, Honduras, Japan, Korea, Malaysia, Mexico, Netherlands, New Zealand, Philippines, Romania, Russian Federation, South Africa, Switzerland, Taiwan, USA, and Uruguay.

In September 2000, up to $50 million worth of food products were recalled due to contamination with Starlink genetically modified corn, which had been approved only for animal consumption and had not been approved for human consumption, and was subsequently withdrawn from the market.

Origin

A *Tripsacum* grass (big) and a teosinte (small)

Maize is the domesticated variant of teosinte. The two plants have dissimilar appearance, maize having a single tall stalk with multiple leaves and teosinte being a short, bushy plant. The difference between the two is largely controlled by differences in just two genes.

Several theories had been proposed about the specific origin of maize in Mesoamerica:

1. It is a direct domestication of a Mexican annual teosinte, *Zea mays* ssp. *parviglumis*, native to the Balsas River valley in south-eastern Mexico, with up to 12% of its genetic material obtained from *Zea mays* ssp. *mexicana* through introgression. This theory was further confirmed by the 2002 study of Matsuoka et al.

2. It has been derived from hybridization between a small domesticated maize (a slightly changed form of a wild maize) and a teosinte of section *Luxuriantes*, either *Z. luxurians* or *Z. diploperennis*.

3. It has undergone two or more domestications either of a wild maize or of a teosinte. (The term "teosinte" describes all species and subspecies in the genus *Zea*, excluding *Zea mays* ssp. *mays*.)

4. It has evolved from a hybridization of *Z. diploperennis* by *Tripsacum dactyloides*.

In the late 1930s, Paul Mangelsdorf suggested that domesticated maize was the result of a hybridization event between an unknown wild maize and a species of *Tripsacum*, a related genus. This theory about the origin of maize has been refuted by modern genetic testing, which refutes Mangelsdorf's model and the fourth listed above.

The teosinte origin theory was proposed by the Russian botanist Nikolai Ivanovich Vavilov in 1931 and the later American Nobel Prize-winner George Beadle in 1932.[10] It is supported experimentally and by recent studies of the plants' genomes. Teosinte and maize are able to cross-breed and produce fertile offspring. A number of questions remain concerning the species, among them:

1. how the immense diversity of the species of sect. *Zea* originated,

2. how the tiny archaeological specimens of 3500–2700 BC could have been selected from a teosinte, and

3. how domestication could have proceeded without leaving remains of teosinte or maize with teosintoid traits earlier than the earliest known until recently, dating from ca. 1100 BC.

The domestication of maize is of particular interest to researchers—archaeologists, geneticists, ethnobotanists, geographers, etc. The process is thought by some to have started 7,500 to 12,000 years ago. Research from the 1950s to 1970s originally focused on the hypothesis that maize domestication occurred in the highlands between the states of Oaxaca and Jalisco, because the oldest archaeological remains of maize known at the time were found there.

Connection with 'Parviglumis' Subspecies

Genetic studies led by John Doebley identified *Zea mays* ssp. *parviglumis*, native to the Balsas River valley in Mexico's southwestern highlands, and also known as Balsas teosinte, as being the crop wild relative teosinte genetically most similar to modern maize. This has been confirmed by further more recent studies, which refined this hypothesis somewhat. Archaeobotanical studies published in 2009 now point to the middle part of the Balsas River valley as the more likely location of early domestication; this river is not very long, so these locations are not very distant. Stone milling tools with maize residue have been found in an 8,700-years old layer of deposits in a cave not far from Iguala, Guerrero.

teosinte (top), maize-teosinte hybrid (middle), maize (bottom)

Also, Doebley was part of the team that is credited with first finding, back in 2002, that maize had been domesticated only once, about 9000 years ago, and then spread throughout the Americas.

A primitive corn was being grown in southern Mexico, Central America, and northern South America 7,000 years ago. Archaeological remains of early maize ears, found at Guila Naquitz Cave in the Oaxaca Valley, date back roughly 6,250 years; the oldest ears from caves near Tehuacan, Puebla, date ca. 3,450 BC.

Centeotl, the Aztec deity of maize

Maize pollen dated to 7300 cal B.P. from San Andres, Tabasco, on the Caribbean coast has also been recovered.

As maize was introduced to new cultures, new uses were developed and new varieties selected to better serve in those preparations. Maize was the staple food, or a major staple – along with squash, Andean region potato, quinoa, beans, and amaranth – of most pre-Columbian North American, Mesoamerican, South American, and Caribbean cultures. The Mesoamerican civilization, in particular, was deeply interrelated with maize. Its traditions and rituals involved all aspects of maize cultivation – from the planting to the food preparation. Maize formed the Mesoamerican people's identity.

It is unknown what precipitated its domestication, because the edible portion of the wild variety is too small and hard to obtain to be eaten directly, as each kernel is enclosed in a very hard bivalve shell. It is possible that, early on, teosinte may have been gathered as preferred feed for domestic animals.

Also, back in 1939, George Beadle demonstrated that the kernels of teosinte are readily "popped" for human consumption, like modern popcorn. Some have argued it would have taken too many generations of selective breeding to produce large, compressed ears for efficient cultivation. However, studies of the hybrids readily made by intercrossing teosinte and modern maize suggest this objection is not well founded.

Spreading to the North

Around 2500 BC, maize began to spread to the north; it was first cultivated in what is now the United States at several sites in New Mexico and Arizona, about 2100 BC.

During the first millennium AD, maize cultivation spread more widely in the areas north. In particular, the large-scale adoption of maize agriculture and consumption in eastern North America took place about A.D. 900. Native Americans cleared large forest and grassland areas for the new crop.

In 2005, research by the USDA Forest Service suggested that the rise in maize cultivation 500 to 1,000 years ago in what is now the southeastern United States corresponded with a decline of freshwater mussels, which are very sensitive to environmental changes.

Production

Methods

Seedlings three weeks after sowing

Young stalks

Because it is cold-intolerant, in the temperate zones maize must be planted in the spring. Its root system is generally shallow, so the plant is dependent on soil moisture. As a C4 plant (a plant that uses C4 carbon fixation), maize is a considerably more water-efficient crop than C3 plants (plants that use C3 carbon fixation) like the small grains, alfalfa and soybeans. Maize is most sensitive to drought at the time of silk emergence, when the flowers are ready for pollination. In the United States, a good harvest was traditionally predicted if the maize were "knee-high by the Fourth of July", although modern hybrids generally exceed this growth rate. Maize used for silage is harvested while the plant is green and the fruit immature. Sweet corn is harvested in the "milk stage", after pollination but before starch has formed, between late summer and early to mid-autumn. Field maize is left in the field very late in the autumn to thoroughly dry the grain, and may, in fact, sometimes not be harvested until winter or even early spring. The importance of sufficient soil moisture is shown in many parts of Africa, where periodic drought regularly causes maize crop failure and consequent famine. Although it is grown mainly in wet, hot climates, it has been said to thrive in cold, hot, dry or wet conditions, meaning that it is an extremely versatile crop.

Mature plants showing ears

Maize was planted by the Native Americans in hills, in a complex system known to some as the Three Sisters. Maize provided support for beans, and the beans provided nitrogen derived from nitrogen-fixing rhizobia bacteria which live on the roots of beans and other legumes; and squashes provided ground cover to stop weeds and inhibit evaporation by providing shade over the soil. This method was replaced by single species hill planting where each hill 60–120 cm (2.0–3.9 ft) apart was planted with three or four seeds, a method still used by home gardeners. A later technique was "checked maize", where hills were placed 40 in (1.0 m) apart in each direction, allowing cultivators to run through the field in two directions. In more arid lands, this was altered and seeds were planted in the bottom of 10–12 cm (3.9–4.7 in) deep furrows to collect water. Modern technique plants maize in rows which allows for cultivation while the plant is young, although the hill technique is still used in the maize fields of some Native American reservations. When maize is planted in rows, it also allows for planting of other crops between these rows to make more efficient use of land space.

In North America, fields are often planted in a two-crop rotation with a nitrogen-fixing crop, often alfalfa in cooler climates and soybeans in regions with longer summers. Sometimes a third crop, winter wheat, is added to the rotation.

Many of the maize varieties grown in the United States and Canada are hybrids. Often the varieties have been genetically modified to tolerate glyphosate or to provide protection against natural pests. Glyphosate is an herbicide which kills all plants except those with genetic tolerance. This genetic tolerance is very rarely found in nature.

In midwestern United States, low-till or no-till farming techniques are usually used. In low-till, fields are covered once, maybe twice, with a tillage implement either ahead of crop planting or after the previous harvest. The fields are planted and fertilized. Weeds are controlled through the use of herbicides, and no cultivation tillage is done during the growing season. This technique reduces moisture evaporation from the soil, and thus provides more moisture for the crop. The technologies mentioned in the previous paragraph enable low-till and no-till farming. Weeds compete with the crop for moisture and nutrients, making them undesirable.

Mature field maize ears

Before World War II, most maize in North America was harvested by hand. This involves a large numbers of workers and associated social events (husking or shucking bees). Some one- and two-row mechanical pickers were in use, but the maize combine was not adopted until after the War. By hand or mechanical picker, the entire ear is harvested, which then requires a separate operation of a maize sheller to remove the kernels from the ear. Whole ears of maize were often stored in corn cribs, and these whole ears are a sufficient form for some livestock feeding use. Few modern farms store maize in this manner. Most harvest the grain from the field and store it in bins. The combine with a corn head (with points and snap rolls instead of a reel) does not cut the stalk; it simply pulls the stalk down. The stalk continues downward and is crumpled into a mangled pile on the ground. The ear of maize is too large to pass between slots in a plate as the snap rolls pull the stalk away, leaving only the ear and husk to enter the machinery. The combine separates out the husk and the cob, keeping only the kernels.

For storing grain in bins, the moisture of the grain must be sufficiently low to avoid spoiling. If the moisture content of the harvested grain is too high, grain dryers are used to reduce the moisture content by blowing heated air through the grain. This can require large amounts of energy in the form of combustible gases (propane or natural gas) and electricity to power the blowers.

Quantity

Maize is widely cultivated throughout the world, and a greater weight of maize is produced each year than any other grain. The United States produces 40% of the world's harvest; other top producing countries include China, Brazil, Mexico, Indonesia, India, France and Argentina. Worldwide production was 817 million tonnes in 2009—more than rice (678 million tonnes) or wheat (682 million tonnes). In 2009, over 159 million hectares (390 million acres) of maize were planted worldwide, with a yield of over 5 tonnes per hectare (80 bu/acre). Production can be significantly higher in certain regions of the world; 2009 forecasts for production in Iowa were 11614 kg/ha (185 bu/acre).[Note 1] There is conflicting evidence to support the hypothesis that maize yield potential has increased over the past few decades. This suggests that changes in yield potential are associated with leaf angle, lodging resistance, tolerance of high plant density, disease/pest tolerance, and other agronomic traits rather than increase of yield potential per individual plant.

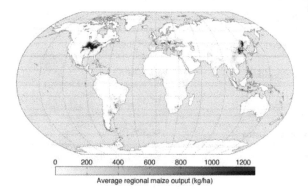

Worldwide maize production

Harvesting maize during the record 2009 season in Jones County, Iowa

Maize in Beijing, China, October 2012

Top ten maize producers in 2013	
Country	**Production (tonnes)**
United States	353,699,441
China	217,730,000
Brazil	80,516,571
Argentina	32,119,211
Ukraine	30,949,550
India	23,290,000
Mexico	22,663,953
Indonesia	18,511,853
France	15,053,100
South Africa	12,365,000
World	**1,016,431,783**

United States

In 2010, the maize planted area for all purposes in the US was estimated at 35 million hectares (87.9 million acres), following an increasing trend since 2008. About 14% of the harvested corn area is irrigated.

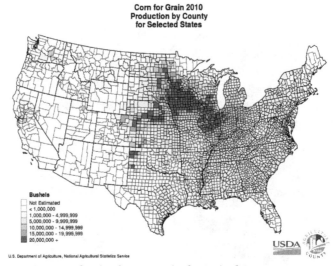

Corn production by county in the United States, 2010

Pests

Insects

- African armyworm (*Spodoptera exempta*)

- Common armyworm (*Pseudaletia unipuncta*)

- Common earwig (*Forficula auricularia*)

- Corn delphacid (*Peregrinus maidis*)

- Corn leaf aphid (*Rhopalosiphum maidis*)

- Corn rootworms (*Diabrotica spp*) including Western corn rootworm (*Diabrotica virgifera virgifera* LeConte), Northern corn rootworm (*D. barberi*) and Southern corn rootworm (*D. undecimpunctata howardi*)

- Corn silkfly (*Euxesta stigmatias*)

- European corn borer (*Ostrinia nubilalis*) (ECB)

- Fall armyworm (*Spodoptera frugiperda*)

- Corn earworm/Cotton bollworm (*Helicoverpa zea*)

- Lesser cornstalk borer (*Elasmopalpus lignosellus*)

- Maize weevil (*Sitophilus zeamais*)

- Northern armyworm, Oriental armyworm or Rice ear-cutting caterpillar (*Mythimna separata*)

- Southwestern corn borer (*Diatraea grandiosella*)

- Stalk borer (*Papaipema nebris*)

The susceptibility of maize to the European corn borer and corn rootworms, and the resulting large crop losses which are estimated at a billion dollars worldwide for each pest, led to the development of transgenics expressing the *Bacillus thuringiensis* toxin. "Bt maize" is widely grown in the United States and has been approved for release in Europe.

Diseases

- Rust

- Corn smut or common smut (*Ustilago maydis*): a fungal disease, known in Mexico as *huitlacoche*, which is prized by some as a gourmet delicacy in itself

- Northern corn leaf blight (Purdue Extension site) (Pioneer site)

- Southern corn leaf blight

- Maize dwarf mosaic virus

- Maize streak virus

- Stewart's Wilt (*Pantoea stewartii*)

- Goss's Wilt (*Clavibacter michiganensis*)

- Grey leaf spot

- Mal de Río Cuarto virus (MRCV)

- Stalk rot

- Ear rot

Uses

Human Food

Vegetable maize (sweet corn)

Maize being roasted over an open flame in India.

Maize and cornmeal (ground dried maize) constitute a staple food in many regions of the world.

Maize is central to Mexican food. Virtually every dish in Mexican cuisine uses maize. In the form of grain or cornmeal, maize is the main ingredient of tortillas, tamales, pozole, atole and all the dishes based on them, like tacos, quesadillas, chilaquiles, enchiladas, tostadas and many more. In Mexico even a fungus of maize, known as huitlacoche is considered a delicacy.

Introduced into Africa by the Portuguese in the 16th century, maize has become Africa's most important staple food crop. Maize meal is made into a thick porridge in many cultures: from the polenta of Italy, the *angu* of Brazil, the *mămăligă* of Romania, to cornmeal mush in the US (and hominy grits in the South) or the food called mealie pap in South Africa and *sadza*, *nshima* and *ugali* in other parts of Africa. Maize meal is also used as a replacement for wheat flour, to make cornbread and other baked products. Masa (cornmeal treated with limewater) is the main ingredient for tortillas, atole and many other dishes of Central American food.

Cut white sweet corn

Popcorn consists of kernels of certain varieties that explode when heated, forming fluffy pieces that are eaten as a snack. Roasted dried maize ears with semihardened kernels, coated with a seasoning mixture of fried chopped spring onions with salt added to the oil, is a popular snack food in Vietnam. *Cancha*, which are roasted maize chulpe kernels, are a very popular snack food in Peru, and also appears in traditional Peruvian *ceviche*. An unleavened bread called *makki di roti* is a popular bread eaten in the Punjab region of India and Pakistan.

Dried maize *mote*, also known as hominy, is used in Mexican cuisine

Chicha and *chicha morada* (purple chicha) are drinks typically made from particular types of maize. The first one is fermented and alcoholic, the second is a soft drink commonly drunk in Peru.

Corn flakes are a common breakfast cereal in North America and the United Kingdom, and found in many other countries all over the world.

Maize can also be prepared as hominy, in which the kernels are soaked with lye in a process called nixtamalization; or grits, which are coarsely ground hominy. These are commonly eaten in the

Southeastern United States, foods handed down from Native Americans, who called the dish sagamite.

The Brazilian dessert *canjica* is made by boiling maize kernels in sweetened milk. Maize can also be harvested and consumed in the unripe state, when the kernels are fully grown but still soft. Unripe maize must usually be cooked to become palatable; this may be done by simply boiling or roasting the whole ears and eating the kernels right off the cob. Sweet corn, a genetic variety that is high in sugars and low in starch, is usually consumed in the unripe state. Such corn on the cob is a common dish in the United States, Canada, United Kingdom, Cyprus, some parts of South America, and the Balkans, but virtually unheard of in some European countries. Corn on the cob was hawked on the streets of early 19th-century New York City by poor, barefoot "Hot Corn Girls", who were thus the precursors of hot dog carts, churro wagons, and fruit stands seen on the streets of big cities today. The cooked, unripe kernels may also be shaved off the cob and served as a vegetable in side dishes, salads, garnishes, etc. Alternatively, the raw unripe kernels may also be grated off the cobs and processed into a variety of cooked dishes, such as maize purée, tamales, *pamonhas*, *curau*, cakes, ice creams, etc.

Maize is a major source of starch. Cornstarch (maize flour) is a major ingredient in home cooking and in many industrialized food products. Maize is also a major source of cooking oil (corn oil) and of maize gluten. Maize starch can be hydrolyzed and enzymatically treated to produce syrups, particularly high fructose corn syrup, a sweetener; and also fermented and distilled to produce grain alcohol. Grain alcohol from maize is traditionally the source of Bourbon whiskey. Maize is sometimes used as the starch source for beer. Within the United States, the usage of maize for human consumption constitutes about 1/40th of the amount grown in the country. In the United States and Canada, maize is mostly grown to feed livestock, as forage, silage (made by fermentation of chopped green cornstalks), or grain. Maize meal is also a significant ingredient of some commercial animal food products, such as dog food.

Nutritional Value

In a 100-gram serving, maize kernels provide 86 calories and are a good source (10-19% of the Daily Value) of the B vitamins, thiamin, niacin, pantothenic acid (B5) and folate (right table for raw, uncooked kernels, USDA Nutrient Database). In moderate amounts, they also supply dietary fiber and the essential minerals, magnesium and phosphorus whereas other nutrients are in low amounts.

Chemicals

Starch from maize can also be made into plastics, fabrics, adhesives, and many other chemical products.

The corn steep liquor, a plentiful watery byproduct of maize wet milling process, is widely used in the biochemical industry and research as a culture medium to grow many kinds of microorganisms.

Chrysanthemin is found in purple corn and is used as a food coloring.

Bio-fuel

"Feed maize" is being used increasingly for heating; specialized corn stoves (similar to wood stoves) are available and use either feed maize or wood pellets to generate heat. Maize cobs are also used as a biomass fuel source. Maize is relatively cheap and home-heating furnaces have been developed which use maize kernels as a fuel. They feature a large hopper that feeds the uniformly sized maize kernels (or wood pellets or cherry pits) into the fire.

Maize is increasingly used as a feedstock for the production of ethanol fuel. Ethanol is mixed with gasoline to decrease the amount of pollutants emitted when used to fuel motor vehicles. High fuel prices in mid-2007 led to higher demand for ethanol, which in turn led to higher prices paid to farmers for maize. This led to the 2007 harvest being one of the most profitable maize crops in modern history for farmers. Because of the relationship between fuel and maize, prices paid for the crop now tend to track the price of oil.

The price of food is affected to a certain degree by the use of maize for biofuel production. The cost of transportation, production, and marketing are a large portion (80%) of the price of food in the United States. Higher energy costs affect these costs, especially transportation. The increase in food prices the consumer has been seeing is mainly due to the higher energy cost. The effect of biofuel production on other food crop prices is indirect. Use of maize for biofuel production increases the demand, and therefore price of maize. This, in turn, results in farm acreage being diverted from other food crops to maize production. This reduces the supply of the other food crops and increases their prices.

Farm-based maize silage digester located near Neumünster in Germany, 2007. Green inflatable biogas holder is shown on top of the digester

Maize is widely used in Germany as a feedstock for biogas plants. Here the maize is harvested, shredded then placed in silage clamps from which it is fed into the biogas plants. This process makes use of the whole plant rather than simply using the kernels as in the production of fuel ethanol.

A biomass gasification power plant in Strem near Güssing, Burgenland, Austria, began in 2005. Research is being done to make diesel out of the biogas by the Fischer Tropsch method.

Increasingly, ethanol is being used at low concentrations (10% or less) as an additive in gasoline (gasohol) for motor fuels to increase the octane rating, lower pollutants, and reduce petroleum use (what is nowadays also known as "biofuels" and has been generating an intense debate regarding the human beings' necessity of new sources of energy, on the one hand, and the need to maintain, in regions such as Latin America, the food habits and culture which has been the essence of civilizations such as the one originated in Mesoamerica; the entry, January 2008, of maize among the commercial agreements of NAFTA has increased this debate, considering the bad labor conditions of workers in the fields, and mainly the fact that NAFTA "opened the doors to the import of maize from the United States, where the farmers who grow it receive multimillion dollar subsidies and other government supports. (...) According to OXFAM UK, after NAFTA went into effect, the price of maize in Mexico fell 70% between 1994 and 2001. The number of farm jobs dropped as well: from 8.1 million in 1993 to 6.8 million in 2002. Many of those who found themselves without work were small-scale maize growers."). However, introduction in the northern latitudes of the US of tropical maize for biofuels, and not for human or animal consumption, may potentially alleviate this.

As a result of the US federal government announcing its production target of 35 billion US gallons (130,000,000 m³) of biofuels by 2017, ethanol production will grow to 7 billion US gallons (26,000,000 m³) by 2010, up from 4.5 billion in 2006, boosting ethanol's share of maize demand in the US from 22.6 percent to 36.1 percent.

Ornamental and other uses

Some forms of the plant are occasionally grown for ornamental use in the garden. For this purpose, variegated and colored leaf forms as well as those with colorful ears are used.

Corncobs can be hollowed out and treated to make inexpensive smoking pipes, first manufactured in the United States in 1869.

Children playing in a maize kernel box

An unusual use for maize is to create a "corn maze" (or "maize maze") as a tourist attraction. The idea of a maize maze was introduced by the American Maze Company who created a maze in Pennsylvania in 1993. Traditional mazes are most commonly grown using yew hedges, but these take several years to mature. The rapid growth of a field of maize allows a maze to be laid out using GPS at the start of a growing season and for the maize to grow tall enough to obstruct a visitor's line of sight by the start of the summer. In Canada and the US, these are popular in many farming communities.

Maize kernels can be used in place of sand in a sandboxlike enclosure for children's play.

Stigmas from female maize flowers, popularly called corn silk, are sold as herbal supplements.

Maize is used as a fish bait, called "dough balls". It is particularly popular in Europe for coarse fishing.

Additionally, feed corn is sometimes used by hunters to bait animals such as deer or wild hogs.

Fodder

Maize produces a greater quantity of biomass than other cereal plants, which is used for fodder. Digestibility and palatability are higher when ensiled and fermented, rather than dried.

Commodity

Maize is bought and sold by investors and price speculators as a tradable commodity using corn futures contracts. These "futures" are traded on the Chicago Board of Trade (CBOT) under ticker symbol C. They are delivered every year in March, May, July, September, and December.

United States usage Breakdown

The breakdown of usage of the 12.1-billion-bushel (307-million-tonne) 2008 US maize crop was as follows, according to the World Agricultural Supply and Demand Estimates Report by the USDA.

Use	Amount		
	million bushels	million tonnes	percentage
livestock feed	5,250	133.4	43.4
ethanol production	3,650	92.7	30.2
exports	1,850	47.0	15.3
production of starch, corn oil, sweeteners (HFCS, etc.)	943	24.0	7.8
human consumption—grits, corn flour, corn meal, beverage alcohol	327	8.3	2.7

In the US since 2009/2010, maize feedstock use for ethanol production has somewhat exceeded direct use for livestock feed; maize use for fuel ethanol was 5,130 million bushels (130 million tonnes) in the 2013/2014 marketing year.

A fraction of the maize feedstock dry matter used for ethanol production is usefully recovered as DDGS (dried distillers grains with solubles). In the 2010/2011 marketing year, about 29.1 million tonnes of DDGS were fed to US livestock and poultry. Because starch utilization in fermentation for ethanol production leaves other grain constituents more concentrated in the residue, the feed value per kg of DDGS, with regard to ruminant-metabolizable energy and protein, exceeds that of the grain. Feed value for monogastric animals, such as swine and poultry, is somewhat lower than for ruminants.

Comparison to other Staple Foods

Nutrient contents in %DV of common foods (raw, uncooked) per 100 g

Food	Protein DV	Protein Q	Fiber DV	A	B1	B2	B3	B5	B6	B9	B12	Ch.	C	D	E	K	Ca	Fe	Mg	P	K	Na	Zn	Cu	Mn	Se
																				Minerals						
cooking Re-duction %				10	30	20	25		25	35	0	0	30				10	15	20	10	20	5	10	25		
Corn	20	55	6	1	13	4	16	4	19	19	0	0	0	0	0	1	1	11	31	34	15	1	20	10	42	0
Rice	14	71	1.3	0	12	3	11	20	5	2	0	0	0	0	0	0	1	9	6	7	2	0	8	9	49	22
Wheat	27	51	40	0	28	7	34	19	21	11	0	0	0	0	0	0	3	20	36	51	12	0	28	28	151	128
Soybean	73	132	0	31	58	51	8	8	19	94	0	24	10	0	4	59	28	87	70	70	51	0	33	83	126	25
Pigeon pea	43	91	1	50	43	11	15	13	13	114	0	0	0	0	0	0	13	29	46	37	40	1	18	53	90	12
Potato	4	112	7.3	0	5	2	5	3	15	4	0	0	33	0	0	2	1	4	6	6	12	0	2	5	8	0
Sweet potato	3	82	10	284	5	4	3	8	10	3	0	0	4	0	1	2	3	3	6	5	10	2	2	8	13	1
Spinach	6	119	7.3	188	5	11	4	1	10	49	0	4.5	47	0	10	604	10	15	20	5	16	3	4	6	45	1
Dill	7	32	7	154	4	17	8	4	9	38	0	0	142	0	0	0	21	37	14	7	21	3	6	7	63	0
Carrots	2	24	9.3	334	4	3	5	3	7	5	0	0	10	0	3	16	3	2	3	4	9	3	2	2	7	0
Guava	5	24	18	12	4	2	5	5	6	12	0	0	381	0	4	3	2	1	5	4	12	0	2	11	8	1
Papaya	1	7	5.6	22	2	2	2	2	1	10	0	0	103	0	4	3	2	1	2	1	7	0	0	1	1	1
Pumpkin	2	56	1.6	184	3	6	3	3	3	4	0	0	15	0	5	1	2	2	3	4	10	0	2	6	6	0
Sunflower oil	0		0	0	0	0	0	0	0	0	0	0	0	0	205	7	0	0	0	0	0	0	0	0	0	0
Egg	25	136	0	10	5	28	0	14	7	12	22	45	0	9	5	0	5	10	3	19	4	6	7	5	2	45
Milk	6	138	0	2	3	11	1	4	2	1	7	2.6	0	0	0	0	11	0	2	9	4	2	3	1	0	5

Ch. = Choline; Ca = Calcium; Fe = Iron; Mg = Magnesium; P = Phosphorus; K = Potassium; Na = Sodium; Zn = Zinc; Cu = Copper; Mn = Manganese; Se = Selenium; %DV = % daily value i.e. % of DRI (Dietary Reference Intake) Note: All nutrient values including protein and fiber are in %DV per 100 grams of the food item. Significant values are highlighted in light Gray color and bold letters. Cooking reduction = % Maximum typical reduction in nutrients due to boiling without draining for ovo-lacto-vegetables group Q = Quality of Protein in terms of completeness without adjusting for digestability.

The following table shows the nutrient content of maize and major staple foods in a raw harvested form. Raw forms are not edible and cannot be digested. These must be sprouted, or prepared and cooked for human consumption. In sprouted or cooked form, the relative nutritional and anti-nutritional contents of each of these staples are different from that of raw form of these staples reported in the table below.

Nutrient content of major staple foods

STAPLE:	RDA	Maize / Corn[A]	Rice (white)[B]	Rice (brown)[I]	Wheat[C]	Potato[D]	Cassava[E]	Soybean (Green)[F]	Sweet potato[G]	Sorghum[H]	Yam[Y]	Plantain[Z]
Component (per 100g portion)	Amount	Amount	Amount	Amount	Amount	Amount	Amount	Amount	Amount	Amount	Amount	Amount
Water (g)	3000	10	12	10	13	79	60	68	77	9	70	65
Energy (kJ)		1528	1528	1549	1369	322	670	615	360	1419	494	511
Protein (g)	50	9.4	7.1	7.9	12.6	2.0	1.4	13.0	1.6	11.3	1.5	1.3
Fat (g)		4.74	0.66	2.92	1.54	0.09	0.28	6.8	0.05	3.3	0.17	0.37
Carbohydrates (g)	130	74	80	77	71	17	38	11	20	75	28	32
Fiber (g)	30	7.3	1.3	3.5	12.2	2.2	1.8	4.2	3	6.3	4.1	2.3
Sugar (g)		0.64	0.12	0.85	0.41	0.78	1.7	0	4.18	0	0.5	15
Calcium (mg)	1000	7	28	23	29	12	16	197	30	28	17	3
Iron (mg)	8	2.71	0.8	1.47	3.19	0.78	0.27	3.55	0.61	4.4	0.54	0.6
Magnesium (mg)	400	127	25	143	126	23	21	65	25	0	21	37
Phosphorus (mg)	700	210	115	333	288	57	27	194	47	287	55	34
Potassium (mg)	4700	287	115	223	363	421	271	620	337	350	816	499
Sodium (mg)	1500	35	5	7	2	6	14	15	55	6	9	4
Zinc (mg)	11	2.21	1.09	2.02	2.65	0.29	0.34	0.99	0.3	0	0.24	0.14
Copper (mg)	0.9	0.31	0.22		0.43	0.11	0.10	0.13	0.15	-	0.18	0.08

	RDI	corn, yellow [A]	rice, white [B]	rice, brown [I]	wheat [C]	potato [D]	cassava [E]	soybeans [F]	sweet potato [G]	sorghum [H]	yam [Y]	plantains [Z]
Manganese (mg)	2,3	0.49	1.09	3.74	3.99	0.15	0.38	0.55	0.26	-	0.40	-
Selenium (µg)	55	15.5	15.1		70.7	0.3	0.7	1.5	0.6	0	0.7	1.5
Vitamin C (mg)	90	0	0	0	0	19.7	20.6	29	2.4	0	17.1	18.4
Thiamin (B1)(mg)	1.2	0.39	0.07	0.40	0.30	0.08	0.09	0.44	0.08	0.24	0.11	0.05
Riboflavin (B2)(mg)	1.3	0.20	0.05	0.09	0.12	0.03	0.05	0.18	0.06	0.14	0.03	0.05
Niacin (B3) (mg)	16	3.63	1.6	5.09	5.46	1.05	0.85	1.65	0.56	2.93	0.55	0.69
Pantothenic acid (B5) (mg)	5	0.42	1.01	1.49	0.95	0.30	0.11	0.15	0.80	-	0.31	0.26
Vitamin B6 (mg)	1.3	0.62	0.16	0.51	0.3	0.30	0.09	0.07	0.21	-	0.29	0.30
Folate Total (B9) (µg)	400	19	8	20	38	16	27	165	11	0	23	22
Vitamin A (IU)	5000	214	0	0	9	2	13	180	14187	0	138	1127
Vitamin E, alpha-tocopherol (mg)	15	0.49	0.11	0.59	1.01	0.01	0.19	0	0.26	0	0.39	0.14
Vitamin K1 (µg)	120	0.3	0.1	1.9	1.9	1.9	1.9	0	1.8	0	2.6	0.7
Beta-carotene (µg)	10500	97	0		5	1	8	0	8509	0	83	457
Lutein+zeaxanthin (µg)		1355	0		220	8	0	0	0	0	0	30
Saturated fatty acids (g)		0.67	0.18	0.58	0.26	0.03	0.07	0.79	0.02	0.46	0.04	0.14
Monounsaturated fatty acids (g)		1.25	0.21	1.05	0.2	0.00	0.08	1.28	0.00	0.99	0.01	0.03
Polyunsaturated fatty acids (g)		2.16	0.18	1.04	0.63	0.04	0.05	3.20	0.01	1.37	0.08	0.07

A corn, yellow
B rice, white, long-grain, regular, raw, unenriched
C wheat, hard red winter
D potato, flesh and skin, raw
E cassava, raw
F soybeans, green, raw
G sweet potato, raw, unprepared
H sorghum, raw
Y yam, raw
Z plantains, raw
I rice, brown, long-grain, raw

Hazards

Pellagra

When maize was first introduced into farming systems other than those used by traditional native-American peoples, it was generally welcomed with enthusiasm for its productivity. However, a widespread problem of malnutrition soon arose wherever maize was introduced as a staple food. This was a mystery, since these types of malnutrition were not normally seen among the indigenous Americans, for whom maize was the principal staple food.

It was eventually discovered that the indigenous Americans had learned to soak maize in alkali-water—made with ashes and lime (calcium oxide) since at least 1200–1500 BC by Mesoamericans and North Americans—which liberates the B-vitamin niacin, the lack of which was the underlying cause of the condition known as pellagra.

Maize was introduced into the diet of nonindigenous Americans without the necessary cultural knowledge acquired over thousands of years in the Americas. In the late 19th century, pellagra reached epidemic proportions in parts of the southern US, as medical researchers debated two theories for its origin: the deficiency theory (which was eventually shown to be true) said that pellagra was due to a deficiency of some nutrient, and the germ theory said that pellagra was caused by a germ transmitted by stable flies. A third theory, promoted by the eugenicist Charles Davenport, held that people only contracted pellagra if they were susceptible to it due to certain "constitutional, inheritable" traits of the affected individual.

Once alkali processing and dietary variety were understood and applied, pellagra disappeared in the developed world. The development of high lysine maize and the promotion of a more balanced diet have also contributed to its demise. Pellagra still exists today in food-poor areas and refugee camps where people survive on donated maize.

Allergy

Maize contains lipid transfer protein, an indigestible protein that survives cooking. This protein has been linked to a rare and understudied allergy to maize in humans. The allergic reaction can cause skin rash, swelling or itching of mucous membranes, diarrhea, vomiting, asthma and, in severe cases, anaphylaxis. It is unclear how common this allergy is in the general population.

Art

Maize has been an essential crop in the Andes since the pre-Columbian era. The Moche culture from Northern Peru made ceramics from earth, water, and fire. This pottery was a sacred substance, formed in significant shapes and used to represent important themes. Maize represented anthropomorphically as well as naturally.

In the United States, maize ears along with tobacco leaves are carved into the capitals of columns in the United States Capitol building. Maize itself is sometimes used for temporary architectural detailing when the intent is to celebrate the fall season, local agricultural productivity and culture. Bundles of dried maize stalks are often displayed often along with pumpkins, gourds and straw in autumnal displays outside homes and businesses. A well-known example of architectural use is the Corn Palace in Mitchell, South Dakota, which uses cobs and ears of colored maize to implement a mural design that is recycled annually.

Gold maize. Moche culture 300 A.D., Larco Museum, Lima, Peru

A maize stalk with two ripe ears is depicted on the reverse of the Croatian 1 lipa coin, minted since 1993.

Water tower in Rochester, Minnesota being painted as an ear of maize

Rabi Crop

Wheat

Rabi crops or Rabi harvest are agricultural crops sown in winter and harvested in the spring in the South Asia. The term is derived from the Arabic word for "spring", which is used in the Indian subcontinent, where it is the spring harvest (also known as the "winter crop").

Barley

The rabi crops are sown around mid-November, after the monsoon rains are over, and harvesting begins in April/May. The crops are grown either with rainwater that has percolated into the ground, or with irrigation. A good rain in winter spoils the rabi crops but is good for kharif crops.

The major rabi crop in India is wheat, followed by barley, mustard, sesame and peas. Peas are harvested early, as they are ready early: Indian markets are flooded with green peas from January to March, peaking in February.

Many crops are cultivated in both kharif and rabi seasons. The agriculture crops produced in India are seasonal in nature and highly dependent on these two monsoons.

Examples of Rabi Crops:

Cereals

- wheat (*Triticum aestvium*)

- oat (*Avena sativa*)

- barley

- maize (*Zea mays, L.*)

Seed plants

- alfalfa (Lucerne, *Medicago sativa*)

- linseed

- sesame

- cumin (*Cuminum cyminum, L*)

- coriander (*Coriandrum sativum, L*)

- mustard (*Brassica juncea L.*)

- fennel (*Foeniculum vulgare*)

- fenugreek (*Trigonella foenumgraecum, L*)

- isabgol (*Plantago ovata*)

Vegetables

- pea

- chickpea (Gram, *Cicer arientinum*)

- onion (*Allium cepa, L.*)

- tomato (*Solanum lycopersicum, L*)

- potato (*Solanum tuberosum*)

Examples Of Rabi Crop

Wheat

Wheat (*Triticum* spp.) is a cereal grain, (botanically, a type of fruit called a caryopsis) originally from the Levant region of the Near East but now cultivated worldwide. In 2013, world production of wheat was 713 million tons, making it the third most-produced cereal after maize (1,016 million tons) and rice (745 million tons). Wheat was the second most-produced cereal in 2009; world production in that year was 682 million tons, after maize (817 million tons), and with rice as a close third (679 million tons).

This grain is grown on more land area than any other commercial food. World trade in wheat is greater than for all other crops combined. Globally, wheat is the leading source of vegetable protein in human food, having a higher protein content than the other major cereals maize (corn) and rice. In terms of total production tonnages used for food, it is currently second to rice as the main human food crop and ahead of maize, after allowing for maize's more extensive use in animal feeds. The archaeological record suggests that this first occurred in the regions known as the Fertile Crescent.

Origin

Spikelets of a hulled wheat, einkorn

Cultivation and repeated harvesting and sowing of the grains of wild grasses led to the creation of domestic strains, as mutant forms ('sports') of wheat were preferentially chosen by farmers. In domesticated wheat, grains are larger, and the seeds (inside the spikelets) remain attached to the ear by a toughened rachis during harvesting. In wild strains, a more fragile rachis allows the ear to easily shatter and disperse the spikelets. Selection for these traits by farmers might not have been deliberately intended, but simply have occurred because these traits made gathering the seeds easier; nevertheless such 'incidental' selection was an important part of crop domestication. As the traits that improve wheat as a food source *also* involve the loss of the plant's natural seed dispersal mechanisms, highly domesticated strains of wheat cannot survive in the wild.

Cultivation of wheat began to spread beyond the Fertile Crescent after about 8000 BCE. Jared Diamond traces the spread of cultivated emmer wheat starting in the Fertile Crescent sometime before 8800 BCE. Archaeological analysis of wild *emmer* indicates that it was first cultivated in the southern Levant with finds dating back as far as 9600 BCE. Genetic analysis of wild *einkorn* wheat suggests that it was first grown in the Karacadag Mountains in southeastern Turkey. Dated archeological remains of einkorn wheat in settlement sites near this region, including those at Abu Hureyra in Syria, suggest the domestication of einkorn near the Karacadag Mountain Range. With the anomalous exception of two grains from Iraq ed-Dubb, the earliest carbon-14 date for einkorn wheat remains at Abu Hureyra is 7800 to 7500 years BCE.

Remains of harvested emmer from several sites near the Karacadag Range have been dated to between 8600 (at Cayonu) and 8400 BCE (Abu Hureyra), that is, in the Neolithic period. With the exception of Iraq ed-Dubb, the earliest carbon-14 dated remains of domesticated emmer wheat were found in the earliest levels of Tell Aswad, in the Damascus basin, near Mount Hermon in Syria. These remains were dated by Willem van Zeist and his assistant Johanna Bakker-Heeres to 8800 BCE. They also concluded that the settlers of Tell Aswad did not develop this form of emmer themselves, but brought the domesticated grains with them from an as yet unidentified location elsewhere.

The cultivation of emmer reached Greece, Cyprus and India by 6500 BCE, Egypt shortly after 6000 BCE, and Germany and Spain by 5000 BCE. "The early Egyptians were developers of bread and the use of the oven and developed baking into one of the first large-scale food production industries." By 3000 BCE, wheat had reached the British Isles and Scandinavia. A millennium later it reached China. The first identifiable bread wheat (*Triticum aestivum*) with sufficient gluten for yeasted breads has been identified using DNA analysis in samples from a granary dating to approximately 1350 BCE at Assiros in Greek Macedonia.

From Asia, wheat continued to spread throughout Europe. In the British Isles, wheat straw (thatch) was used for roofing in the Bronze Age, and was in common use until the late 19th century.

Farming Techniques

Technological advances in soil preparation and seed placement at planting time, use of crop rotation and fertilizers to improve plant growth, and advances in harvesting methods have all combined to promote wheat as a viable crop. Agricultural cultivation using horse collar leveraged plows (at about 3000 BCE) was one of the first innovations that increased productivity. Much later, when the use of seed drills replaced broadcasting sowing of seed in the 18th century, another great increase in productivity occurred.

Green wheat a month before harvest

Wheat harvest on the Palouse, Idaho, United States

Young wheat crop in a field near Solapur, Maharashtra, India

Yields of pure wheat per unit area increased as methods of crop rotation were applied to long cultivated land, and the use of fertilizers became widespread. Improved agricultural husbandry has more recently included threshing machines and reaping machines (the 'combine harvester'), tractor-drawn cultivators and planters, and better varieties. Great expansion of wheat production occurred as new arable land was farmed in the Americas and Australia in the 19th and 20th centuries.

Genetics

Wheat genetics is more complicated than that of most other domesticated species. Some wheat species are diploid, with two sets of chromosomes, but many are stable polyploids, with four sets of chromosomes (tetraploid) or six (hexaploid).

- Einkorn wheat (*T. monococcum*) is diploid (AA, two complements of seven chromosomes, 2n=14).

- Most tetraploid wheats (e.g. emmer and durum wheat) are derived from wild emmer, *T. dicoccoides*. Wild emmer is itself the result of a hybridization between two diploid wild grasses, *T. urartu* and a wild goatgrass such as *Aegilops searsii* or *Ae. speltoides*. The unknown grass has never been identified among now surviving wild grasses, but the closest living relative is *Aegilops speltoides*. The hybridization that formed wild emmer (AABB) occurred in the wild, long before domestication, and was driven by natural selection.

- Hexaploid wheats evolved in farmers' fields. Either domesticated emmer or durum wheat hybridized with yet another wild diploid grass (*Aegilops tauschii*) to make the hexaploid wheats, spelt wheat and bread wheat. These have *three* sets of paired chromosomes, three times as many as in diploid wheat.

The presence of certain versions of wheat genes has been important for crop yields. Apart from mutant versions of genes selected in antiquity during domestication, there has been more recent deliberate selection of alleles that affect growth characteristics. Genes for the 'dwarfing' trait, first used by Japanese wheat breeders to produce short-stalked wheat, have had a huge effect on wheat yields world-wide, and were major factors in the success of the Green Revolution in Mexico and Asia, an initiative led by Norman Borlaug. Dwarfing genes enable the carbon that is fixed in the plant during photosynthesis to be diverted towards seed production, and they also help prevent the problem of lodging. 'Lodging' occurs when an ear stalk falls over in the wind and rots on the ground, and heavy nitrogenous fertilization of wheat makes the grass grow taller and become more susceptible to this problem. By 1997, 81% of the developing world's wheat area was planted to semi-dwarf wheats, giving both increased yields and better response to nitrogenous fertilizer.

Wild grasses in the genus *Triticum* and related genera, and grasses such as rye have been a source of many disease-resistance traits for cultivated wheat breeding since the 1930s.

Heterosis, or hybrid vigor (as in the familiar F1 hybrids of maize), occurs in common (hexaploid) wheat, but it is difficult to produce seed of hybrid cultivars on a commercial scale (as is done with maize) because wheat flowers are perfect and normally self-pollinate. Commercial hybrid wheat seed has been produced using chemical hybridizing agents; these chemicals selectively interfere with pollen development, or naturally occurring cytoplasmic male sterility systems. Hybrid wheat has been a limited commercial success in Europe (particularly France), the United States and South Africa. F1 hybrid wheat cultivars should not be confused with the standard method of breeding inbred wheat cultivars by crossing two lines using hand emasculation, then selfing or inbreeding the progeny many (ten or more) generations before release selections are identified to be released as a variety or cultivar.

Synthetic hexaploids made by crossing the wild goatgrass wheat ancestor *Aegilops tauschii* and various durum wheats are now being deployed, and these increase the genetic diversity of cultivated wheats.

Stomata (or leaf pores) are involved in both uptake of carbon dioxide gas from the atmosphere and water vapor losses from the leaf due to water transpiration. Basic physiological investigation of these gas exchange processes has yielded valuable carbon isotope based methods that are used for breeding wheat varieties with improved water-use efficiency. These varieties can improve crop productivity in rain-fed dry-land wheat farms.

In 2010, a team of UK scientists funded by BBSRC announced they had decoded the wheat genome for the first time (95% of the genome of a variety of wheat known as Chinese Spring line 42). This genome was released in a basic format for scientists and plant breeders to use but was not a fully annotated sequence which was reported in some of the media.

On 29 November 2012, an essentially complete gene set of bread wheat has been published. Random shotgun libraries of total DNA and cDNA from the *T. aestivum* cv. Chinese Spring (CS42) were sequenced in Roche 454 pyrosequencer using GS FLX Titanium and GS FLX+ platforms to generate 85 Gb of sequence (220 million reads), equivalent to 5X genome coverage and identified between 94,000 and 96,000 genes.

This sequence data provides direct access to about 96,000 genes, relying on orthologous gene sets from other cereals. and represents an essential step towards a systematic understanding of biology and engineering the cereal crop for valuable traits. Its implications in cereal genetics and breeding includes the examination of genome variation, association mapping using natural populations, performing wide crosses and alien introgression, studying the expression and nucleotide polymorphism in transcriptomes, analyzing population genetics and evolutionary biology, and studying the epigenetic modifications. Moreover, the availability of large-scale genetic markers generated through NGS technology will facilitate trait mapping and make marker-assisted breeding much feasible.

Moreover, the data not only facilitate in deciphering the complex phenomena such as heterosis and epigenetics, it may also enable breeders to predict which fragment of a chromosome is derived from which parent in the progeny line, thereby recognizing crossover events occurring in every progeny line and inserting markers on genetic and physical maps without ambiguity. In due course, this will assist in introducing specific chromosomal segments from one cultivar to another. Besides, the researchers had identified diverse classes of genes participating in energy production, metabolism and growth that were probably linked with crop yield, which can now be utilized for the development of transgenic wheat. Thus whole genome sequence of wheat and the availability of thousands of SNPs will inevitably permit the breeders to stride towards identifying novel traits, providing biological knowledge and empowering biodiversity-based breeding.

Plant Breeding

Sheaved and stooked wheat

In traditional agricultural systems wheat populations often consist of landraces, informal farmer-maintained populations that often maintain high levels of morphological diversity. Although landraces of wheat are no longer grown in Europe and North America, they continue to be important elsewhere. The origins of formal wheat breeding lie in the nineteenth century, when single line varieties were created through selection of seed from a single plant noted to have desired properties. Modern wheat breeding developed in the first years of the twentieth century and was closely linked to the development of Mendelian genetics. The standard method of breeding inbred wheat cultivars is by crossing two lines using hand emasculation, then selfing or inbreeding the progeny. Selections are *identified* (shown to have the genes responsible for the varietal differences) ten or more generations before release as a variety or cultivar.

Wheat

The major breeding objectives include high grain yield, good quality, disease and insect resistance and tolerance to abiotic stresses, including mineral, moisture and heat tolerance. The major diseases in temperate environments include the following, arranged in a rough order of their significance from cooler to warmer climates: eyespot, Stagonospora nodorum blotch (also known as glume blotch), yellow or stripe rust, powdery mildew, Septoria tritici blotch (sometimes known as leaf blotch), brown or leaf rust, Fusarium head blight, tan spot and stem rust. In tropical areas, spot blotch (also known as Helminthosporium leaf blight) is also important.

Wheat has also been the subject of mutation breeding, with the use of gamma, x-rays, ultraviolet light, and sometimes harsh chemicals. The varieties of wheat created through these methods are in the hundreds (going as far back as 1960), more of them being created in higher populated countries such as China. Bread wheat with high grain iron and zinc content was developed through gamma radiation breeding.

Hybrid Wheat

Because wheat self-pollinates, creating hybrid varieties is extremely labor-intensive; the high cost of hybrid wheat seed relative to its moderate benefits have kept farmers from adopting them widely despite nearly 90 years of effort. F1 hybrid wheat cultivars should not be confused with wheat cultivars deriving from standard plant breeding. Heterosis or hybrid vigor (as in the familiar F1 hybrids of maize) occurs in common (hexaploid) wheat, but it is difficult to produce seed of hybrid cultivars on a commercial scale as is done with maize because wheat flowers are perfect in the botanical sense, meaning they have both male and female parts, and normally self-pollinate. Commercial hybrid wheat seed has been produced using chemical hybridizing agents, plant growth regulators that selectively interfere with pollen development, or naturally occurring cytoplasmic male sterility systems. Hybrid wheat has been a limited commercial success in Europe (particularly France), the United States and South Africa.

Hulled Versus Free-threshing Wheat

The four wild species of wheat, along with the domesticated varieties einkorn, emmer and spelt, have hulls. This more primitive morphology (in evolutionary terms) consists of toughened glumes that tightly enclose the grains, and (in domesticated wheats) a semi-brittle rachis that breaks easily on threshing. The result is that when threshed, the wheat ear breaks up into spikelets. To obtain the grain, further processing, such as milling or pounding, is needed to remove the hulls or husks. In contrast, in free-threshing (or naked) forms such as durum wheat and common wheat, the glumes are fragile and the rachis tough. On threshing, the chaff breaks up, releasing the grains. Hulled wheats are often stored as spikelets because the toughened glumes give good protection against pests of stored grain.

Left: Naked wheat, Bread wheat *Triticum aestivum*; Right: Hulled wheat, Einkorn, *Triticum monococcum*.
Note how the einkorn ear breaks down into intact spikelets.

Naming

There are many botanical classification systems used for wheat species, discussed in a separate article on wheat taxonomy. The name of a wheat species from one information source may not be the name of a wheat species in another.

Within a species, wheat cultivars are further classified by wheat breeders and farmers in terms of:

- Growing season, such as winter wheat vs. spring wheat.

- Protein content. Bread wheat protein content ranges from 10% in some soft wheats with high starch contents, to 15% in hard wheats.

- The quality of the wheat protein gluten. This protein can determine the suitability of a wheat to a particular dish. A strong and elastic gluten present in bread wheats enables dough to trap carbon dioxide during leavening, but elastic gluten interferes with the rolling of pasta into thin sheets. The gluten protein in durum wheats used for pasta is strong but not elastic.

- Grain color (red, white or amber). Many wheat varieties are reddish-brown due to phenolic compounds present in the bran layer which are transformed to pigments by browning enzymes. White wheats have a lower content of phenolics and browning enzymes, and are generally less astringent in taste than red wheats. The yellowish color of durum wheat and semolina flour made from it is due to a carotenoid pigment called lutein, which can be oxidized to a colorless form by enzymes present in the grain.

Sack of wheat

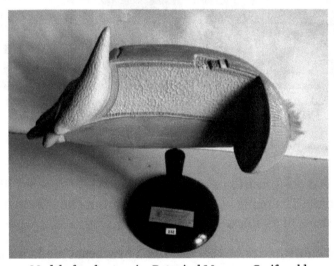

Model of a wheat grain, Botanical Museum Greifswald

Major Cultivated Species of Wheat

Hexaploid species

- Common wheat or bread wheat (*T. aestivum*) – A hexaploid species that is the most widely cultivated in the world.

- Spelt (*T. spelta*) – Another hexaploid species cultivated in limited quantities. Spelt is sometimes considered a subspecies of the closely related species common wheat (*T. aestivum*), in which case its botanical name is considered to be *Triticum aestivum* subsp. *spelta*.

Tetraploid species

- Durum (*T. durum*) – The only tetraploid form of wheat widely used today, and the second most widely cultivated wheat.

- Emmer (*T. dicoccon*) – A tetraploid species, cultivated in ancient times but no longer in widespread use.

- Khorasan (Triticum turgidum ssp. turanicum also called Triticum turanicum) is a tetraploid wheat species. It is an ancient grain type; Khorasan refers to a historical region in modern-day Afghanistan and the northeast of Iran. This grain is twice the size of modern-day wheat and is known for its rich nutty flavor.

Diploid species

- Einkorn (*T. monococcum*) – A diploid species with wild and cultivated variants. Domesticated at the same time as emmer wheat, but never reached the same importance.

Classes used in the United States:

- Durum – Very hard, translucent, light-colored grain used to make semolina flour for pasta & bulghur; high in protein, specifically, gluten protein.

- Hard Red Spring – Hard, brownish, high-protein wheat used for bread and hard baked goods. Bread Flour and high-gluten flours are commonly made from hard red spring wheat. It is primarily traded at the Minneapolis Grain Exchange.

- Hard Red Winter – Hard, brownish, mellow high-protein wheat used for bread, hard baked goods and as an adjunct in other flours to increase protein in pastry flour for pie crusts. Some brands of unbleached all-purpose flours are commonly made from hard red winter wheat alone. It is primarily traded on the Kansas City Board of Trade. One variety is known as "turkey red wheat", and was brought to Kansas by Mennonite immigrants from Russia.

- Soft Red Winter – Soft, low-protein wheat used for cakes, pie crusts, biscuits, and muffins. Cake flour, pastry flour, and some self-rising flours with baking powder and salt added, for example, are made from soft red winter wheat. It is primarily traded on the Chicago Board of Trade.

- Hard White – Hard, light-colored, opaque, chalky, medium-protein wheat planted in dry, temperate areas. Used for bread and brewing.

- Soft White – Soft, light-colored, very low protein wheat grown in temperate moist areas. Used for pie crusts and pastry. Pastry flour, for example, is sometimes made from soft white winter wheat.

Red wheats may need bleaching; therefore, white wheats usually command higher prices than red wheats on the commodities market.

As a Food

Raw wheat can be ground into flour or, using hard durum wheat only, can be ground into semolina; germinated and dried creating malt; crushed or cut into cracked wheat; parboiled (or steamed), dried, crushed and de-branned into bulgur also known as groats. If the raw wheat is broken into parts at the mill, as is usually done, the outer husk or bran can be used several ways. Wheat is a major ingredient in such foods as bread, porridge, crackers, biscuits, Muesli, pancakes, pies, pastries, cakes, cookies, muffins, rolls, doughnuts, gravy, boza (a fermented beverage), and breakfast cereals (e.g., Wheatena, Cream of Wheat, Shredded Wheat, and Wheaties).

Wheat is used in a wide variety of foods.

Nutrition

In 100 grams, wheat provides 327 calories and is an excellent source (more than 19% of the Daily Value, DV) of multiple essential nutrients, such as protein, dietary fiber, manganese, phosphorus and niacin (table). Several B vitamins and other dietary minerals are in significant content (table). Wheat is 13% water, 71% carbohydrates, 1.5% fat and 13% protein (table).

Nutrient contents in %DV of common foods (raw, uncooked) per 100 g

Food	Protein DV	Protein Q	Fiber DV	A	B1	B2	B3	B5	B6	B9	B12	Ch.	C	D	E	K	Ca	Fe	Mg	P	K	Na	Zn	Cu	Mn	Se
cooking Reduction %				10	30	20	25		25	35	0	0	30				10	15	20	10	20	5	10	25		
Corn	20	55	6	1	13	4	16	4	19	19	0	0	0	0	0	1	1	11	31	34	15	1	20	10	42	0
Rice	14	71	1.3	0	12	3	11	20	5	2	0	0	0	0	0	0	1	9	6	7	2	0	8	9	49	22
Wheat	27	51	40	0	28	7	34	19	21	11	0	0	0	0	0	0	3	20	36	51	12	0	28	28	151	128
Soybean	73	132	0	31	58	51	8	8	19	94	0	24	10	0	4	59	28	87	70	70	51	0	33	83	126	25
Pigeon pea	43	91	1	50	43	11	15	13	13	114	0	0	0	0	0	0	13	29	46	37	40	1	18	53	90	12
Potato	4	112	7.3	0	5	2	5	3	15	4	0	0	33	0	0	2	1	4	6	6	12	0	2	5	8	0
Sweet potato	3	82	10	284	5	4	3	8	10	3	0	0	4	0	1	2	3	3	6	5	10	2	2	8	13	1
Spinach	6	119	7.3	188	5	11	4	1	10	49	0	4.5	47	0	10	604	10	15	20	5	16	3	4	6	45	1
Dill	7	32	7	154	4	17	8	4	9	38	0	0	142	0	0	0	21	37	14	7	21	3	6	7	63	0
Carrots	2		9.3	334	4	3	5	3	7	5	0	0	10	0	3	16	3	2	3	4	9	3	2	2	7	0
Guava	5	24	18	12	4	2	5	5	6	12	0	0	381	0	4	3	2	1	5	4	12	0	2	11	8	1
Papaya	1	7	5.6	22	2	2	2	2	1	10	0	0	103	0	4	3	2	1	2	1	7	0	0	1	1	1
Pumpkin	2	56	1.6	184	3	6	3	3	3	4	0	0	15	0	5	1	2	4	3	4	10	0	2	6	6	0
Sunflower oil	0		0	0	0	0	0	0	0	0	0	0	0	0	205	7	0	0	0	0	0	0	0	0	0	0
Egg	25	136	0	10	5	28	0	14	7	12	22	45	0	9	5	0	5	10	3	19	4	6	7	5	2	45
Milk	6	138	0	2	3	11	1	4	2	1	7	2.6	0	0	0	0	11	0	2	9	4	2	3	1	0	5

Ch. = Choline; Ca = Calcium; Fe = Iron; Mg = Magnesium; P = Phosphorus; K = Potassium; Na = Sodium; Zn = Zinc; Cu = Copper; Mn = Manganese; Se = Selenium; %DV = % daily value i.e. % of DRI (Dietary Reference Intake) Note: All nutrient values including protein and fiber are in %DV per 100 grams of the food item. Significant values are highlighted in light Gray color and bold letters.

Cooking reduction = % Maximum typical reduction in nutrients due to boiling without draining for ovo-lacto-vegetables group Q = Quality of Protein in terms of completeness without adjusting for digestability.

100 g (3.5 oz) of hard red winter wheat contain about 12.6 g (0.44 oz) of protein, 1.5 g (0.053 oz) of total fat, 71 g (2.5 oz) of carbohydrate (by difference), 12.2 g (0.43 oz) of dietary fiber, and 3.2 mg (0.00011 oz) of iron (17% of the daily requirement); the same weight of hard red spring wheat contains about 15.4 g (0.54 oz) of protein, 1.9 g (0.067 oz) of total fat, 68 g (2.4 oz) of carbohydrate (by difference), 12.2 g (0.43 oz) of dietary fiber, and 3.6 mg (0.00013 oz) of iron (20% of the daily requirement).

Worldwide Consumption

Wheat is grown on more than 218,000,000 hectares (540,000,000 acres), larger than for any other crop. World trade in wheat is greater than for all other crops combined. With rice, wheat is the world's most favored staple food. It is a major diet component because of the wheat plant's agronomic adaptability with the ability to grow from near arctic regions to equator, from sea level to plains of Tibet, approximately 4,000 m (13,000 ft) above sea level. In addition to agronomic adaptability, wheat offers ease of grain storage and ease of converting grain into flour for making edible, palatable, interesting and satisfying foods. Wheat is the most important source of carbohydrate in a majority of countries.

Wheat protein is easily digested by nearly 99% of the human population (all but those with gluten-related disorders), as is its starch. With a small amount of animal or legume protein added, a wheat-based meal is highly nutritious.

The most common forms of wheat are white and red wheat. However, other natural forms of wheat exist. Other commercially minor but nutritionally promising species of naturally evolved wheat species include black, yellow and blue wheat.

Health Concerns

Coeliac disease affects 1-2% of the general population, but most cases remain unrecognized, undiagnosed and untreated.

While coeliac disease is caused by a reaction to wheat proteins, it is not the same as a wheat allergy. Other diseases triggered by gluten consumption are non-celiac gluten sensitivity, (estimated in one study to affect the general population in a wide range from 0.5% to 13%), gluten ataxia and dermatitis herpetiformis.

Comparison of Wheat with other Major Staple Foods

The following table shows the nutrient content of wheat and other major staple foods in a raw form.

Raw forms of these staples, however, are not edible and cannot be digested. These must be sprouted, or prepared and cooked as appropriate for human consumption. In sprouted or cooked form, the relative nutritional and anti-nutritional contents of each of these grains is remarkably different from that of raw form of these grains reported in this table.

In cooked form, the nutrition value for each staple depends on the cooking method (for example: baking, boiling, steaming, frying, etc.).

Nutrient content of major staple foods

STAPLE: Component (per 100g portion)	RDA Amount	Maize / Corn[A] Amount	Rice (white)[B] Amount	Rice (brown)[I] Amount	Wheat[C] Amount	Potato[D] Amount	Cassava[E] Amount	Soybean (Green)[F] Amount	Sweet potato[G] Amount	Sorghum[H] Amount	Yam[Y] Amount	Plantain[Z] Amount
Water (g)	3000	10	12	10	13	79	60	68	77	9	70	65
Energy (kJ)		1528	1528	1549	1369	322	670	615	360	1419	494	511
Protein (g)	50	9.4	7.1	7.9	12.6	2.0	1.4	13.0	1.6	11.3	1.5	1.3
Fat (g)		4.74	0.66	2.92	1.54	0.09	0.28	6.8	0.05	3.3	0.17	0.37
Carbohydrates (g)	130	74	80	77	71	17	38	11	20	75	28	32
Fiber (g)	30	7.3	1.3	3.5	12.2	2.2	1.8	4.2	3	6.3	4.1	2.3
Sugar (g)		0.64	0.12	0.85	0.41	0.78	1.7	0	4.18	0	0.5	15
Calcium (mg)	1000	7	28	23	29	12	16	197	30	28	17	3
Iron (mg)	8	2.71	0.8	1.47	3.19	0.78	0.27	3.55	0.61	4.4	0.54	0.6
Magnesium (mg)	400	127	25	143	126	23	21	65	25	0	21	37
Phosphorus (mg)	700	210	115	333	288	57	27	194	47	287	55	34
Potassium (mg)	4700	287	115	223	363	421	271	620	337	350	816	499
Sodium (mg)	1500	35	5	7	2	6	14	15	55	6	9	4
Zinc (mg)	11	2.21	1.09	2.02	2.65	0.29	0.34	0.99	0.3	0	0.24	0.14
Copper (mg)	0.9	0.31	0.22		0.43	0.11	0.10	0.13	0.15	-	0.18	0.08

Nutrient	RDA	A	B	I	C	D	E	F	G	H	Y	Z
Manganese (mg)	2.3	0.49	1.09	3.74	3.99	0.15	0.38	0.55	0.26	–	0.40	–
Selenium (µg)	55	15.5	15.1		70.7	0.3	0.7	1.5	0.6	0	0.7	1.5
Vitamin C (mg)	90	0	0	0	0	19.7	20.6	29	2.4	0	17.1	18.4
Thiamin (B1)(mg)	1.2	0.39	0.07	0.40	0.30	0.08	0.09	0.44	0.08	0.24	0.11	0.05
Riboflavin (B2)(mg)	1.3	0.20	0.05	0.09	0.12	0.03	0.05	0.18	0.06	0.14	0.03	0.05
Niacin (B3) (mg)	16	3.63	1.6	5.09	5.46	1.05	0.85	1.65	0.56	2.93	0.55	0.69
Pantothenic acid (B5) (mg)	5	0.42	1.01	1.49	0.95	0.30	0.11	0.15	0.80	–	0.31	0.26
Vitamin B6 (mg)	1.3	0.62	0.16	0.51	0.3	0.30	0.09	0.07	0.21	–	0.29	0.30
Folate Total (B9) (µg)	400	19	8	20	38	16	27	165	11	0	23	22
Vitamin A (IU)	5000	214	0	0	9	2	13	180	14187	0	138	1127
Vitamin E, alpha-tocopherol (mg)	15	0.49	0.11	0.59	1.01	0.01	0.19	0	0.26	0	0.39	0.14
Vitamin K1 (µg)	120	0.3	0.1	1.9	1.9	1.9	1.9	0	1.8	0	2.6	0.7
Beta-carotene (µg)	10500	97	0		5	1	8	0	8509	0	83	457
Lutein+zeaxanthin (µg)		1355	0		220	8	0	0	0	0	0	30
Saturated fatty acids (g)		0.67	0.18	0.58	0.26	0.03	0.07	0.79	0.02	0.46	0.04	0.14
Monounsaturated fatty acids (g)		1.25	0.21	1.05	0.2	0.00	0.08	1.28	0.00	0.99	0.01	0.03
Polyunsaturated fatty acids (g)		2.16	0.18	1.04	0.63	0.04	0.05	3.20	0.01	1.37	0.08	0.07

A corn, yellow
B rice, white, long-grain, regular, raw, unenriched
C wheat, hard red winter
D potato, flesh and skin, raw
E cassava, raw
F soybeans, green, raw
G sweet potato, raw, unprepared
H sorghum, raw
Y yam, raw
Z plantains, raw
I rice, brown, long-grain, raw

Commercial use

Harvested wheat grain that enters trade is classified according to grain properties for the purposes of the commodity markets. Wheat buyers use these to decide which wheat to buy, as each class has special uses, and producers use them to decide which classes of wheat will be most profitable to cultivate.

Wheat is widely cultivated as a cash crop because it produces a good yield per unit area, grows well in a temperate climate even with a moderately short growing season, and yields a versatile, high-quality flour that is widely used in baking. Most breads are made with wheat flour, including many breads named for the other grains they contain, for example, most rye and oat breads. The popularity of foods made from wheat flour creates a large demand for the grain, even in economies with significant food surpluses.

Utensil made of dry wheat branches for loaves of bread

In recent years, low international wheat prices have often encouraged farmers in the United States to change to more profitable crops. In 1998, the price at harvest was $2.68 per bushel. USDA report revealed that in 1998, average operating costs were $1.43 per bushel and total costs were $3.97 per bushel. In that study, farm wheat yields averaged 41.7 bushels per acre (2.2435 metric ton/hectare), and typical total wheat production value was $31,900 per farm, with total farm production value (including other crops) of $173,681 per farm, plus $17,402 in government payments. There were significant profitability differences between low- and high-cost farms, mainly due to crop yield differences, location, and farm size.

In 2007 there was a dramatic rise in the price of wheat due to freezes and flooding in the Northern Hemisphere and a drought in Australia. Wheat futures in September, 2007 for December and March delivery had risen above $9.00 a bushel, prices never seen before. There were complaints in Italy about the high price of pasta.

Production and Consumption

In 2011, global per capita wheat consumption was 65 kg (143 lb), with the highest per capita consumption of 210 kg (460 lb) found in Azerbaijan. In 1997, global wheat consumption was 101 kg (223 lb) per capita, with the highest consumption 623 kg (1,373 lb) per capita in Denmark, but most of this (81%) was for animal feed. Wheat is the primary food staple in North Africa and the Middle East, and is growing in popularity in Asia. Unlike rice, wheat production is more widespread globally though China's share is almost one-sixth of the world.

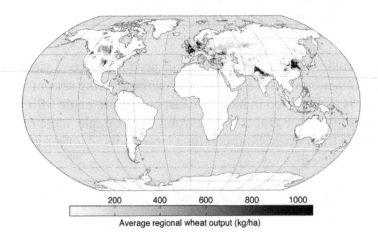

A map of worldwide wheat production.

The combine Claas Lexion 584 06833 is threshing the wheat. The combine crushes the chaff and blows it across the field.

"There is a little increase in yearly crop yield comparison to the year 1990. The reason for this is not in development of sowing area, but the slow and successive increasing of the average yield. Average 2.5 tons wheat was produced on one hectare crop land in the world in the first half of 1990s, however this value was about 3 tons in 2009. In the world per capita wheat producing area continuously decreased between 1990 and 2009 considering the change of world population. There was no significant change in wheat producing area in this period. However, due to the improvement of average yields there is some fluctuation in each year considering the per capita production, but there is no considerable decline. In 1990 per capita production was 111.98 kg/capita/year, while it was already 100.62 kg/capita/year in 2009. The decline is evident and the per capita production level of the year 1990 can not be feasible simultaneously with the growth of world population in spite of the increased average yields. In the whole period the lowest per capita production was in 2006."

The combine Claas Lexion 584 06833 mows, threshes, shreddes the chaff and blows it across the field. At the same time the combine loads the threshed wheat onto a trailer while moving at full speed.

In the 20th century, global wheat output expanded by about 5-fold, but until about 1955 most of this reflected increases in wheat crop area, with lesser (about 20%) increases in crop yields per unit area. After 1955 however, there was a ten-fold increase in the rate of wheat yield improvement per year, and this became the major factor allowing global wheat production to increase. Thus technological innovation and scientific crop management with synthetic nitrogen fertilizer, irrigation and wheat breeding were the main drivers of wheat output growth in the second half of the century. There were some significant decreases in wheat crop area, for instance in North America.

Better seed storage and germination ability (and hence a smaller requirement to retain harvested crop for next year's seed) is another 20th century technological innovation. In Medieval England, farmers saved one-quarter of their wheat harvest as seed for the next crop, leaving only three-quarters for food and feed consumption. By 1999, the global average seed use of wheat was about 6% of output.

Several factors are currently slowing the rate of global expansion of wheat production: population growth rates are falling while wheat yields continue to rise, and the better economic profitability of other crops such as soybeans and maize, linked with investment in modern genetic technologies, has promoted shifts to other crops.

Farming Systems

In the Punjab region of India and Pakistan, as well as North China, irrigation has been a major contributor to increased grain output. More widely over the last 40 years, a massive increase in fertilizer use together with the increased availability of semi-dwarf varieties in developing countries, has greatly increased yields per hectare. In developing countries, use of (mainly nitrogenous) fertilizer increased 25-fold in this period. However, farming systems rely on much more than fertilizer and breeding to improve productivity. A good illustration of this is Australian wheat growing in the southern winter cropping zone, where, despite low rainfall (300 mm), wheat cropping is successful even with relatively little use of nitrogenous fertilizer. This is achieved by 'rotation cropping' (traditionally called the ley system) with leguminous pastures and, in the last decade, including a canola crop in the rotations has boosted wheat yields by a further 25%. In these low rainfall areas, better use of available soil-water (and better control of soil erosion) is achieved by retaining the stubble after harvesting and by minimizing tillage.

Woman harvesting wheat, Raisen district, Madhya Pradesh, India

In 2009, the most productive farms for wheat were in France producing 7.45 metric tonnes per hectare (although French production has low protein content and requires blending with higher protein wheat to meet the specifications required in some countries). The five largest producers of wheat in 2009 were China (115 million metric tonnes), India (81 MMT), Russian Federation (62 MMT), United States (60 MMT) and France (38 MMT). The wheat farm productivity in India and Russia were about 35% of the wheat farm productivity in France. China's farm productivity for wheat, in 2009, was about double that of Russia.

In addition to gaps in farming system technology and knowledge, some large wheat grain producing countries have significant losses after harvest at the farm and because of poor roads, inadequate storage technologies, inefficient supply chains and farmers' inability to bring the produce into retail markets dominated by small shopkeepers. Various studies in India, for example, have concluded that about 10% of total wheat production is lost at farm level, another 10% is lost because of poor storage and road networks, and additional amounts lost at the retail level. One study claims that if these post-harvest wheat grain losses could be eliminated with better infrastructure and retail network, in India alone enough food would be saved every year to feed 70 to 100 million people over a year.

Futures Contracts

Wheat futures are traded on the Chicago Board of Trade, Kansas City Board of Trade, and Minneapolis Grain Exchange, and have delivery dates in March (H), May (K), July (N), September (U), and December (Z).

Geographical Variation

Top wheat producers (in million metric tons)					
Rank	Country	2010	2011	2012	2013
1	China	115	117	126	122
2	India	80	86	95	94

Top wheat producers (in million metric tons)					
Rank	Country	2010	2011	2012	2013
3	United States	60	54	62	58
4	Russia	41	56	38	52
5	France	40	38	40	39
6	Canada	23	25	27	38
7	Germany	24	22	22	25
8	Pakistan	23	25	24	24
9	Australia	22	27	30	23
11	Ukraine	16	22	16	23
10	Turkey	19	21	20	22
12	Iran	13	13	14	14
13	Kazakhstan	9	22	13	14
14	United Kingdom	14	15	13	12
15	Poland	9	9	9	9
—	*World*	651	704	675	713

There are substantial differences in wheat farming, trading, policy, sector growth, and wheat uses in different regions of the world. In the EU and Canada for instance, there is significant addition of wheat to animal feeds, but less so in the United States.

The biggest wheat producer in 2010 was EU-27, followed by China, India, United States and Russian Federation.

The largest exporters of wheat in 2009 were, in order of exported quantities: United States, EU-27, Canada, Russian Federation, Australia, Ukraine and Kazakhstan. Upon the results of 2011, Ukraine became the world's sixth wheat exporter as well. The largest importers of wheat in 2009 were, in order of imported quantities: Egypt, EU-27, Brazil, Indonesia, Algeria and Japan. EU-27 was on both export and import list, because EU countries such as Italy and Spain imported wheat, while other EU-27 countries exported their harvest. The Black Sea region – which includes Kazakhstan, the Russian Federation and Ukraine – is amongst the most promising area for grain exporters; it possess significant production potential in terms of both wheat yield and area increases. The Black Sea region is also located close to the traditional grain importers in the Middle East, North Africa and Central Asia.

In the rapidly developing countries of Asia, westernization of diets associated with increasing prosperity is leading to growth in *per capita* demand for wheat at the expense of the other food staples.

In the past, there has been significant governmental intervention in wheat markets, such as price supports in the US and farm payments in the EU. In the EU these subsidies have encouraged heavy use of fertilizer inputs with resulting high crop yields. In Australia and Argentina direct government subsidies are much lower.

World's Most Productive Wheat Farms and Farmers

The average annual world farm yield for wheat was 3.3 tonnes per hectare (330 grams per square meter), in 2013.

New Zealand wheat farms were the most productive in 2013, with a nationwide average of 9.1 tonnes per hectare. Ireland was a close second.

Various regions of the world hold wheat production yield contests every year. Yields above 12 tonnes per hectare are routinely achieved in many parts of the world. Chris Dennison of Oamaru, New Zealand, set a world record for wheat yield in 2003 at 15.015 tonnes per hectare (223 bushels/acre). In 2010, this record was surpassed by another New Zealand farmer, Michael Solari, with 15.636 tonnes per hectare (232.64 bushels/acre) at Otama, Gore.

Agronomy

Wheat spikelet with the three anthers sticking out

Crop Development

Wheat normally needs between 110 and 130 days between sowing and harvest, depending upon climate, seed type, and soil conditions (winter wheat lies dormant during a winter freeze). Optimal crop management requires that the farmer have a detailed understanding of each stage of development in the growing plants. In particular, spring fertilizers, herbicides, fungicides, and growth regulators are typically applied only at specific stages of plant development. For example, it is currently recommended that the second application of nitrogen is best done when the ear (not visible at this stage) is about 1 cm in size (Z31 on Zadoks scale). Knowledge of stages is also important to identify periods of higher risk from the climate. For example, pollen formation from the mother cell, and the stages between anthesis and maturity are susceptible to high temperatures, and this adverse effect is made worse by water stress. Farmers also benefit from knowing when the 'flag leaf' (last leaf) appears, as this leaf represents about 75% of photosynthesis reactions during the grain filling period, and so should be preserved from disease or insect attacks to ensure a good yield.

Several systems exist to identify crop stages, with the Feekes and Zadoks scales being the most widely used. Each scale is a standard system which describes successive stages reached by the crop during the agricultural season.

Wheat at the anthesis stage. Face view (left) and side view (right) and wheat ear at the late milk

Diseases

Rust-affected wheat seedlings

There are many wheat diseases, mainly caused by fungi, bacteria, and viruses. Plant breeding to develop new disease-resistant varieties, and sound crop management practices are important for preventing disease. Fungicides, used to prevent the significant crop losses from fungal disease, can be a significant variable cost in wheat production. Estimates of the amount of wheat production lost owing to plant diseases vary between 10–25% in Missouri. A wide range of organisms infect wheat, of which the most important are viruses and fungi.

The main wheat-disease categories are:

- Seed-borne diseases: these include seed-borne scab, seed-borne *Stagonospora* (previously known as *Septoria*), common bunt (stinking smut), and loose smut. These are managed with fungicides.

- Leaf- and head- blight diseases: Powdery mildew, leaf rust, *Septoria tritici* leaf blotch, *Stagonospora* (*Septoria*) nodorum leaf and glume blotch, and *Fusarium* head scab.

- Crown and root rot diseases: Two of the more important of these are 'take-all' and *Cephalosporium* stripe. Both of these diseases are soil borne.

- Stem rust diseases: Caused by basidiomycete fungi e.g. Ug99

- Viral diseases: Wheat spindle streak mosaic (yellow mosaic) and barley yellow dwarf are the two most common viral diseases. Control can be achieved by using resistant varieties.

Pests

Wheat is used as a food plant by the larvae of some Lepidoptera (butterfly and moth) species including the flame, rustic shoulder-knot, setaceous Hebrew character and turnip moth. Early in the season, many species of birds, including the long-tailed widowbird, and rodents feed upon wheat crops. These animals can cause significant damage to a crop by digging up and eating newly planted seeds or young plants. They can also damage the crop late in the season by eating the grain from the mature spike. Recent post-harvest losses in cereals amount to billions of dollars per year in the United States alone, and damage to wheat by various borers, beetles and weevils is no exception. Rodents can also cause major losses during storage, and in major grain growing regions, field mice numbers can sometimes build up explosively to plague proportions because of the ready availability of food. To reduce the amount of wheat lost to post-harvest pests, Agricultural Research Service scientists have developed an "insect-o-graph," which can detect insects in wheat that are not visible to the naked eye. The device uses electrical signals to detect the insects as the wheat is being milled. The new technology is so precise that it can detect 5-10 infested seeds out of 300,000 good ones. Tracking insect infestations in stored grain is critical for food safety as well as for the marketing value of the crop.

Oat

The oat (*Avena sativa*), sometimes called the common oat, is a species of cereal grain grown for its seed, which is known by the same name (usually in the plural, unlike other cereals and pseudocereals). While oats are suitable for human consumption as oatmeal and rolled oats, one of the most common uses is as livestock feed.

Origin

The wild ancestor of *Avena sativa* and the closely related minor crop, *A. byzantina*, is the hexaploid wild oat *A. sterilis*. Genetic evidence shows the ancestral forms of *A. sterilis* grew in the Fertile Crescent of the Near East. Domesticated oats appear relatively late, and far from the Near East, in Bronze Age Europe. Oats, like rye, are usually considered a secondary crop, i.e., derived from a weed of the primary cereal domesticates wheat and barley. As these cereals spread westwards into cooler, wetter areas, this may have favored the oat weed component, and have led to its domestication.

Cultivation

Top Ten Oats Producers—2013 (Thousand Metric Tons)	
Russia	4,027
Canada	2,680
Poland	1,439
Finland	1,159
Australia	1,050
United States	929
Spain	799
United Kingdom	784

Sweden	*776*
Germany	*668*
World Total	*20,732*

Oats are best grown in temperate regions. They have a lower summer heat requirement and greater tolerance of rain than other cereals, such as wheat, rye or barley, so are particularly important in areas with cool, wet summers, such as Northwest Europe and even Iceland. Oats are an annual plant, and can be planted either in autumn (for late summer harvest) or in the spring (for early autumn harvest).

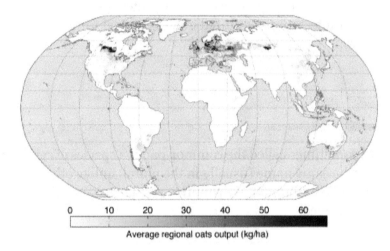

Worldwide oat production

Uses

Closeup of oat florets (small flowers)

Oats have numerous uses in foods; most commonly, they are rolled or crushed into oatmeal, or ground into fine oat flour. Oatmeal is chiefly eaten as porridge, but may also be used in a variety of baked goods, such as oatcakes, oatmeal cookies and oat bread. Oats are also an ingredient in many cold cereals, in particular muesli and granola.

Historical attitudes towards oats have varied. Oat bread was first manufactured in Britain, where the first oat bread factory was established in 1899. In Scotland, they were, and still are, held in high esteem, as a mainstay of the national diet.

In Scotland, a dish was made by soaking the husks from oats for a week, so the fine, floury part of the meal remained as sediment to be strained off, boiled and eaten. Oats are also widely used there as a thickener in soups, as barley or rice might be used in other countries.

Oats are also commonly used as feed for horses when extra carbohydrates and the subsequent boost in energy are required. The oat hull may be crushed ("rolled" or "crimped") for the horse to more easily digest the grain, or may be fed whole. They may be given alone or as part of a blended food pellet. Cattle are also fed oats, either whole or ground into a coarse flour using a roller mill, burr mill, or hammer mill.

Winter oats may be grown as an off-season groundcover and ploughed under in the spring as a green fertilizer, or harvested in early summer. They also can be used for pasture; they can be grazed a while, then allowed to head out for grain production, or grazed continuously until other pastures are ready.

Oat straw is prized by cattle and horse producers as bedding, due to its soft, relatively dust-free, and absorbent nature. The straw can also be used for making corn dollies. Tied in a muslin bag, oat straw was used to soften bath water.

Oats are also occasionally used in several different drinks. In Britain, they are sometimes used for brewing beer. Oatmeal stout is one variety brewed using a percentage of oats for the wort. The more rarely used oat malt is produced by the Thomas Fawcett & Sons Maltings and was used in the Maclay Oat Malt Stout before Maclays Brewery ceased independent brewing operations. A cold, sweet drink called *avena* made of ground oats and milk is a popular refreshment throughout Latin America. Oatmeal caudle, made of ale and oatmeal with spices, was a traditional British drink and a favourite of Oliver Cromwell.

Oat extract can also be used to soothe skin conditions.

Oat grass has been used traditionally for medicinal purposes, including to help balance the menstrual cycle, treat dysmenorrhoea and for osteoporosis and urinary tract infections.

Health

Nutrient Profile

Oats are generally considered healthy due to their rich content of several essential nutrients (table). In a 100 gram serving, oats provide 389 calories and are an excellent source (20% or more of the Daily Value, DV) of protein (34% DV), dietary fiber (44% DV), several B vitamins and numerous dietary minerals, especially manganese (233% DV) (table). Oats are 66% carbohydrates, including 11% dietary fiber and 4% beta-glucans, 7% fat and 17% protein (table).

The established property of their cholesterol-lowering effects has led to acceptance of oats as a health food.

Oat grains in their husks

A sample of oat bran

Soluble Fiber

Oat bran is the outer casing of the oat. Its daily consumption over weeks lowers LDL ("bad") and total cholesterol, possibly reducing the risk of heart disease.

One type of soluble fiber, beta-glucans, has been proven to lower cholesterol.

After reports of research finding that dietary oats can help lower cholesterol, the United States Food and Drug Administration (FDA) issued a final rule that allows food companies to make health claims on food labels of foods that contain soluble fiber from whole oats (oat bran, oat flour and rolled oats), noting that 3.0 grams of soluble fiber daily from these foods may reduce the risk of heart disease. To qualify for the health claim, the whole oat-containing food must provide at least 0.75 grams of soluble fiber per serving.

Beta-D-glucans, usually referred to as beta-glucans, comprise a class of indigestible polysaccharides widely found in nature in sources such as grains, barley, yeast, bacteria, algae and mushrooms. In oats, barley and other cereal grains, they are located primarily in the endosperm cell wall. The oat beta-glucan health claim applies to oat bran, rolled oats, whole oat flour and oatrim, a soluble fraction of alpha-amylase hydrolyzed oat bran or whole oat flour.

Oat beta-glucan is a viscous polysaccharide made up of units of the monosaccharide D-glucose. Oat beta-glucan is composed of mixed-linkage polysaccharides. This means the bonds between the D-glucose or D-glucopyranosyl units are either beta-1, 3 linkages or beta-1, 4 linkages. This type of beta-glucan is also referred to as a mixed-linkage (1→3), (1→4)-beta-D-glucan. The (1→3)-linkages break up the uniform structure of the beta-D-glucan molecule and make it soluble and flexible. In comparison, the indigestible polysaccharide cellulose is also a beta-glucan, but is not soluble because of its (1→4)-beta-D-linkages. The percentages of beta-glucan in the various whole oat products are: oat bran, having from 5.5% to 23.0%; rolled oats, about 4%; and whole oat flour about 4%.

Fat

Oats, after corn (maize), have the highest lipid content of any cereal, e.g., greater than 10% for oats and as high as 17% for some maize cultivars compared to about 2–3% for wheat and most other cereals. The polar lipid content of oats (about 8–17% glycolipid and 10–20% phospholipid or a total of about 33%) is greater than that of other cereals, since much of the lipid fraction is contained within the endosperm.

Protein

Oats are the only cereal containing a globulin or legume-like protein, avenalin, as the major (80%) storage protein. Globulins are characterised by solubility in dilute saline as opposed to the more typical cereal proteins, such as gluten and zein, the prolamines (prolamins). The minor protein of oat is a prolamine, avenin.

Oat protein is nearly equivalent in quality to soy protein, which World Health Organization research has shown to be equal to meat, milk and egg protein. The protein content of the hull-less oat kernel (groat) ranges from 12 to 24%, the highest among cereals.

Celiac Disease

Celiac disease (coeliac disease) is a permanent intolerance to gluten proteins that appears in genetically predisposed people. Gluten is present in wheat, barley, rye and all their species and hybrids and contains hundreds of proteins, with high contents of prolamins.

Oat avenin toxicity in celiac people depends on the oat cultivar consumed because of prolamin genes, protein amino acid sequences, and the immunoreactivities of toxic prolamins which are different among oat varieties. Some cultivars of oat could be a safe part of a gluten-free diet, requiring knowledge of the oat variety used in food products for a gluten-free diet. Oat products may also be cross-contaminated with gluten-containing cereals. Nevertheless, the long-term effects of pure oats consumption are still unclear. Celiac people who choose to consume oats need a more rigorous lifelong follow-up, possibly including periodic performance of intestinal biopsies.

Agronomy

Oats are sown in the spring or early summer in colder areas, as soon as the soil can be worked. An early start is crucial to good fields, as oats go dormant in summer heat. In warmer areas, oats are sown in late summer or early fall. Oats are cold-tolerant and are unaffected by late frosts or snow.

Oats in Saskatchewan near harvest time

Noire d'Epinal, an ancient oat variety.

Seeding Rates

Typically, about 125 to 175 kg/ha (between 2.75 and 3.25 bushels per acre) are sown, either broadcast or drilled. Lower rates are used when interseeding with a legume. Somewhat higher rates can be used on the best soils, or where there are problems with weeds. Excessive sowing rates lead to problems with lodging, and may reduce yields.

Fertilizer Requirements

Oats remove substantial amounts of nitrogen from the soil. They also remove phosphorus in the form of P_2O_5 at the rate of 0.25 pound per bushel per acre (1 bushel = 38 pounds at 12% moisture). Phosphate is thus applied at a rate of 30 to 40 kg/ha, or 30 to 40 lb/acre. Oats remove potash

(K_2O) at a rate of 0.19 pound per bushel per acre, which causes it to use 15–30 kg/ha, or 13–27 lb/acre. Usually, 50–100 kg/ha (45–90 lb/ac) of nitrogen in the form of urea or anhydrous ammonia is sufficient, as oats use about one pound per bushel per acre. A sufficient amount of nitrogen is particularly important for plant height and hence, straw quality and yield. When the prior-year crop was a legume, or where ample manure is applied, nitrogen rates can be reduced somewhat.

Weed Control

The vigorous growth of oats tends to choke out most weeds. A few tall broadleaf weeds, such as ragweed, goosegrass, wild mustard, and buttonweed (velvetleaf), occasionally create a problem, as they complicate harvest and reduce yields. These can be controlled with a modest application of a broadleaf herbicide, such as 2,4-D, while the weeds are still small.

Pests and Diseases

Oats are relatively free from diseases and pests with the exception being leaf diseases, such as leaf rust and stem rust. However, *Puccinia coronata* var. avenae is a pathogen that can greatly reduce crop yields. A few lepidopteran caterpillars feed on the plants—e.g. rustic shoulder-knot and setaceous Hebrew character moths, but these rarely become a major pest.

Harvesting

Harvesting of oats in Jølster, Norway *ca.* 1890
(Photo: Axel Lindahl/Norwegian Museum of Cultural History)

Harvest techniques are a matter of available equipment, local tradition, and priorities. Farmers seeking the highest yield from their crops time their harvest so the kernels have reached 35% moisture, or when the greenest kernels are just turning cream-colour. They then harvest by swathing, cutting the plants at about 10 cm (3.9 in) above ground, and putting the swathed plants into windrows with the grain all oriented the same way. They leave the windrows to dry in the sun for several days before combining them using a pickup header. Finally, they bale the straw.

Oats can also be left standing until completely ripe and then combined with a grain head. This causes greater field losses as the grain falls from the heads, and to harvesting losses, as the grain is threshed out by the reel. Without a draper head, there is also more damage to the straw, since it is not properly oriented as it enters the combine's throat. Overall yield loss is 10–15% compared to proper swathing.

Historical harvest methods involved cutting with a scythe or sickle, and threshing under the feet of cattle. Late 19th- and early 20th-century harvesting was performed using a binder. Oats were gathered into shocks, and then collected and run through a stationary threshing machine.

Storage

After combining, the oats are transported to the farmyard using a grain truck, semi, or road train, where they are augered or conveyed into a bin for storage. Sometimes, when there is not enough bin space, they are augered into portable grain rings, or piled on the ground. Oats can be safely stored at 12-14% moisture; at higher moisture levels, they must be aerated or dried.

Yield and Quality

In the United States, No.1 oats weigh 42 pounds per US bushel (541 kg/m^3); No.3 oats must weigh at least 38 lb/US bu (489 kg/m^3). If over 36 lb/US bu (463 kg/m^3), they are graded as No.4 and oats under 36 lb/US bu (463 kg/m^3) are graded as "light weight".

Oat seeds

In Canada, No.1 oats weigh 42.64 lb/US bu (549 kg/m^3); No.2 oats must weigh 40.18 lb/US bu (517 kg/m^3); No.3 oats must weigh at least 38.54 lb/US bu (496 kg/m^3) and if oats are lighter than 36.08 lb/US bu (464 kg/m^3) they do not make No.4 oats and have no grade.

Note, however, that oats are bought and sold and yields are figured, on the basis of a bushel equal to 32 pounds (14.5 kg or 412 kg/m^3) in the United States and a bushel equal to 34 pounds (15.4 kg or 438 kg/m^3) in Canada. "Bright oats" were sold on the basis of a bushel equal to 48 pounds (21.8 kg or 618 kg/m^3) in the United States.

Yields range from 60 to 80 US bushels per acre (5.2–7.0 m^3/ha) on marginal land, to 100 to 150 US bushels per acre (8.7–13.1 m³/ha) on high-producing land. The average production is 100 bushels per acre, or 3.5 tonnes per hectare.

Straw yields are variable, ranging from one to three tonnes per hectare, mainly due to available nutrients and the variety used (some are short-strawed, meant specifically for straight combining).

Processing

Porridge oats before cooking

Oats processing is a relatively simple process:

Cleaning and Sizing

Upon delivery to the milling plant, chaff, rocks, other grains and other foreign material are removed from the oats.

Dehulling

Centrifugal acceleration is used to separate the outer hull from the inner oat groat. Oats are fed by gravity onto the centre of a horizontally spinning stone, which accelerates them towards the outer ring. Groats and hulls are separated on impact with this ring. The lighter oat hulls are then aspirated away, while the denser oat groats are taken to the next step of processing. Oat hulls can be used as feed, processed further into insoluble oat fibre, or used as a biomass fuel.

Kilning

The unsized oat groats pass through a heat and moisture treatment to balance moisture, but mainly to stabilize them. Oat groats are high in fat (lipids) and once removed from their protective hulls and exposed to air, enzymatic (lipase) activity begins to break down the fat into free fatty acids, ultimately causing an off-flavour or rancidity. Oats begin to show signs of enzymatic rancidity within four days of being dehulled if not stabilized. This process is primarily done in food-grade plants, not in feed-grade plants. Groats are not considered raw if they have gone through this process; the heat disrupts the germ and they cannot sprout.

Sizing of Groats

Many whole oat groats break during the dehulling process, leaving the following types of groats to be sized and separated for further processing: whole oat groats, coarse steel cut groats, steel cut groats, and fine steel cut groats. Groats are sized and separated using screens, shakers and indent screens. After the whole oat groats are separated, the remaining broken groats get sized again into the three groups (coarse, regular, fine), and then stored. "Steel cut" refers to all sized or cut groats. When not enough broken groats are available to size for further processing, whole oat groats are sent to a cutting unit with steel blades that evenly cut groats into the three sizes above.

Final Processing

Three methods are used to make the finished product:

Flaking

This process uses two large smooth or corrugated rolls spinning at the same speed in opposite directions at a controlled distance. Oat flakes, also known as rolled oats, have many different sizes, thicknesses and other characteristics depending on the size of oat groats passed between the rolls. Typically, the three sizes of steel cut oats are used to make instant, baby and quick rolled oats, whereas whole oat groats are used to make regular, medium and thick rolled oats. Oat flakes range in thickness from 0.36 mm to 1.00 mm.

Oat Bran Milling

This process takes the oat groats through several roll stands to flatten and separate the bran from the flour (endosperm). The two separate products (flour and bran) get sifted through a gyrating sifter screen to further separate them. The final products are oat bran and debranned oat flour.

Whole Flour Milling

This process takes oat groats straight to a grinding unit (stone or hammer mill) and then over sifter screens to separate the coarse flour and final whole oat flour. The coarser flour is sent back to the grinding unit until it is ground fine enough to be whole oat flour. This method is used often in India and other countries. In India whole grain flour of oats (jai) used to make Indian bread known as jarobra in Himachal Pradesh.

Oat Flour Preparation at Home

Oat flour can be purchased, but one can grind for small scale use by pulsing rolled oats or old-fashioned (not quick) oats in a food processor or spice mill.

Naming

In Scottish English, oats may be referred to as corn. (In the English language, the major staple grain of the local area is often referred to as "corn". In the US, "corn" originates from "Indian corn" and refers to what others call "maize".)

Oats Futures

Oats futures are traded on the Chicago Board of Trade and have delivery dates in March (H), May (K), July (N), September (U) and December (Z).

Cash Crop

A cotton ball. Cotton is a significant cash crop. According to the National Cotton Council of America, in 2014, China was the world's largest cotton-producing country with an estimated 100,991,000 480-pound bales. India was ranked second at 42,185,000 480-pound bales.

Yerba mate (left, a key ingredient in the beverage known as mate), roasted by the fire, coffee beans (middle) and tea (right) are all used for caffeinated infusions and have cash crop histories.

A cash crop is an agricultural crop which is grown for sale to return a profit. It is typically purchased by parties separate from a farm. The term is used to differentiate marketed crops from subsistence crops, which are those fed to the producer's own livestock or grown as food for the producer's family. In earlier times cash crops were usually only a small (but vital) part of a farm's total yield, while today, especially in developed countries, almost all crops are mainly grown for revenue. In the least developed countries, cash crops are usually crops which attract demand in more developed nations, and hence have some export value.

Prices for major cash crops are set in commodity markets with global scope, with some local variation (termed as "basis") based on freight costs and local supply and demand balance. A consequence of this is that a nation, region, or individual producer relying on such a crop may suffer low prices should a bumper crop elsewhere lead to excess supply on the global markets. This system has been criticized by traditional farmers. Coffee is an example of a product that has been susceptible to significant commodity futures price variations.

Globalization

Issues involving subsidies and trade barriers on such crops have become controversial in discussions of globalization. Many developing countries take the position that the current international trade system is unfair because it has caused tariffs to be lowered in industrial goods while allowing for low tariffs and agricultural subsidies for agricultural goods. This makes it difficult for a developing nation to export its goods overseas, and forces developing nations to compete with imported goods which are exported from developed nations at artificially low prices. The practice of exporting at artificially low prices is known as dumping, and is illegal in most nations. Controversy over this issue led to the collapse of the Cancún trade talks in 2003, when the Group of 22 refused to consider agenda items proposed by the European Union unless the issue of agricultural subsidies was addressed.

Per Climate Zones

Arctic

The Arctic climate is generally not conducive for the cultivation of cash crops. However, one potential cash crop for the Arctic is *Rhodiola rosea*, a hardy plant used as a medicinal herb that grows in the Arctic. There is currently consumer demand for the plant, but the available supply is less than the demand (as of 2011).

Temperate

Cash crops grown in regions with a temperate climate include many cereals (wheat, rye, corn, barley, oats), oil-yielding crops (e.g. rapeseed, mustard seeds), vegetables (e.g. potatoes), tree fruit or top fruit (e.g. apples, cherries) and soft fruit (e.g. strawberries, raspberries).

A tea plantation in the Cameron Highlands in Malaysia

Subtropical

In regions with a subtropical climate, oil-yielding crops (e.g. soybeans) and some vegetables and herbs are the predominant cash crops.

Tropical

In regions with a tropical climate, coffee, cocoa, sugar cane, bananas, oranges, cotton and jute (a soft, shiny vegetable fiber that can be spun into coarse, strong threads), are common cash crops. The oil palm is a tropical palm tree, and the fruit from it is used to make palm oil.

By Continent & Country

Africa

Around 60 percent of African workers are employed in the agricultural sector, with about three-fifths of African farmers being subsistence farmers. For example, in Burkina Faso 85% of its residents (over two million people) are reliant upon cotton production for income, and over half of the country's population lives in poverty. Larger farms tend to grow cash crops such as coffee, tea, cotton, cocoa, fruit and rubber. These farms, typically operated by large corporations, cover tens of square kilometres and employ large numbers of laborers. Subsistence farms provide a source of food and a relatively small income for families, but generally fail to produce enough to make re-investment possible.

Jatropha curcas is a cash crop used to produce biofuel.

The situation in which African nations export crops while a significant amount of people on the continent struggle with hunger has been blamed on developed countries, including the United States, Japan and the European Union. These countries protect their own agricultural sectors, through high import tariffs and offer subsidies to their farmers, which some have contended is leading to the overproduction of commodities such as cotton, grain and milk. The result of this is that the global price of such products is continually reduced until Africans are unable to compete in world markets, except in cash crops that do not grow easily in temperate climates.

Africa has realized significant growth in biofuel plantations, many of which are on lands which were purchased by British companies. *Jatropha curcas* is a cash crop grown for biofuel production in Africa. Some have criticized the practice of raising non-food plants for export while Africa has problems with hunger and food shortages, and some studies have correlated the proliferation of land acquisitions, often for use to grow non-food cash crops with increasing hunger rates in Africa.

Australia

Australia produces significant amounts of lentils. It was estimated in 2010 that Australia would produce approximately 143,000 tons of lentils. Most of Australia's lentil harvest is exported to the Indian subcontinent and the Middle East.

United States

Cash cropping in the United States rose to prominence after the baby boomer generation and the end of World War II. It was seen as a way to feed the large population boom and continues to be the main factor in having an affordable food supply in the United States. According to the 1997 U.S. Census of Agriculture, 90% of the farms in the United States are still owned by families, with an additional 6% owned by a partnership. Cash crop farmers have utilized precision agricultural technologies combined with time-tested practices to produce affordable food. Based upon United States Department of Agriculture (USDA) statistics for 2010, states with the highest fruit production quantities are California, Florida and Washington.

Oranges are a significant U.S. cash crop

List of U.S. Cash Crops

Various potato cultivars

Sliced sugarcane, a significant cash crop in Hawaii

Vietnam

Coconut is a cash crop of Vietnam.

Global Cash Crops

Coconut palms are cultivated in more than 80 countries of the world, with a total production of 61 million tonnes per year. The oil and milk derived from it are commonly used in cooking and frying; coconut oil is also widely used in soaps and cosmetics.

Sustainability of Cash Crops

Approximately 70% of the world's food is produced by 500 million smallholder farmers. For their livelihood they depend on the production of cash crops, basic commodities that are hard to differentiate in the market. The great majority (80%) of the world's farms measure 2 hectares or less. These smallholder farmers are mainly found in developing countries and are often unorganized, illiterate or enjoyed only basic education. Smallholder farmers have little bargaining power and incomes are low, leading to a situation in which they cannot invest much in upscaling their businesses. In general, farmers lack access to agricultural inputs and finance, and do not have enough knowledge on good agricultural and business practices. These high level problems are in many cases threatening the future of agricultural sectors and theories start evolving on how to secure a sustainable future for agriculture. Sustainable market transformations are initiated in which industry leaders work together in a pre-competitive environment to change market conditions. Sustainable intensification focuses on facilitating entrepreneurial farmers. To stimulate farm investment projects on access to finance for agriculture are also popping up. One example is the SCOPE methodology, an assessment tool that measures the management maturity and professionalism of producer organizations as to give financing organizations better insights in the risks involved in financing. Currently agricultural finance is always considered risky and avoided by financial institutions.

Black Market Cash Crops

In the U.S., cannabis has been termed as a cash crop.

Coca, opium poppies and cannabis are significant black market cash crops, the prevalence of which varies. In the United States, cannabis is considered by some to be the most valuable cash crop. In 2006, it was reported in a study by Jon Gettman, a marijuana policy researcher, that in contrast to government figures for legal crops such as corn and wheat and using the study's projections for U.S. cannabis production at that time, cannabis was cited as "the top cash crop in 12 states and among the top three cash crops in 30." The study also estimated cannabis production at the time (in 2006) to be valued at $35.8 billion USD, which exceeded the combined value of corn at $23.3 billion and wheat at $7.5 billion.

Example of Cash Crop

Coffee

Coffee	
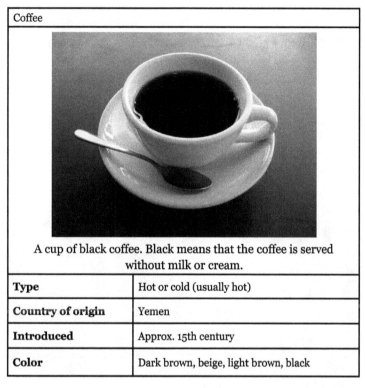 A cup of black coffee. Black means that the coffee is served without milk or cream.	
Type	Hot or cold (usually hot)
Country of origin	Yemen
Introduced	Approx. 15th century
Color	Dark brown, beige, light brown, black

Coffee is a brewed drink prepared from roasted coffee beans, which are the seeds of berries from the *Coffea* plant. The genus Coffea is native to tropical Africa, Madagascar, and the Comoros, Mauritius and Réunion in the Indian Ocean. The plant was exported from Africa to countries around the world and coffee plants are now cultivated in over 70 countries, primarily in the equatorial regions of the Americas, Southeast Asia, India, and Africa. The two most commonly grown are the highly regarded arabica, and the less sophisticated but stronger and more hardy robusta. Once ripe, coffee berries are picked, processed, and dried. Dried coffee seeds (referred to as beans) are roasted to varying degrees, depending on the desired flavor. Roasted beans are ground and brewed with near boiling water to produce coffee as a beverage.

Coffee is slightly acidic and can have a stimulating effect on humans because of its caffeine content. Coffee is one of the most popular drinks in the world. It can be prepared and presented in a variety of ways (e.g., espresso, French press, cafe latte, etc.). It is usually served hot, although iced coffee is also served. Clinical studies indicate that moderate coffee consumption is benign or mildly beneficial in healthy adults, with continuing research on whether long-term consumption inhibits cognitive decline during aging or lowers the risk of some forms of cancer.

The earliest credible evidence of coffee-drinking appears in the middle of the 15th century in the Sufi shrines of Yemen. It was here in Arabia that coffee seeds were first roasted and brewed in a similar way to how it is now prepared. Coffee seeds were first exported from Eastern Africa to Yemen, as the coffee plant is thought to have been indigenous to the former. Yemeni traders took coffee back to their homeland and began to cultivate the seed. By the 16th century, it had reached the rest of the Middle East, Persia, Turkey, and northern Africa. From there, it spread to Europe and the rest of the world.

Coffee is a major export commodity: it is the top agricultural export for numerous countries and is among the world's largest legal agricultural exports. It is one of the most valuable commodities exported by developing countries. Green (unroasted) coffee is one of the most traded agricultural commodities in the world. Some controversy is associated with coffee cultivation and the way developed countries trade with developing nations and the impact of its cultivation on the environment, in regards to clearing of land for coffee-growing and water use. Consequently, fair trade coffee and organic coffee are an expanding market.

Etymology

Coffee beans

The first reference to coffee in the English language is in the form *chaona*, dated to 1598 and understood to be a misprint of *chaoua*, equivalent, in the orthography of the time, to *chaova*. This term and "coffee" both derive from the Ottoman Turkish *kahve*, by way of the Italian *caffè*.

In turn, the Arabic *qahwah* may be an origin, traditionally held to refer to a type of wine whose etymology is given by Arab lexicographers as deriving from the verb *qahiya* (قهي), "to lack hunger", in reference to the drink's reputation as an appetite suppressant. It has also been proposed that the source may be the Proto-Central Semitic root q-h-h meaning "dark".

Alternatively, the word Khat, a plant widely used as stimulant in Yemen and Ethiopia before being supplanted by coffee has been suggested as a possible origin, or the Arabic word *quwwah'* (meaning "strength"). It may also come from the Kingdom of Kaffa in southeast Ethiopia where Coffea arabica grows wild, but this is considered less likely; in the local Kaffa language, the coffee plant is instead called "bunno".

History

Over the door of a Leipzig coffeeshop is a sculptural representation
of a man in Turkish dress, receiving a cup of coffee from a boy

Legendary Accounts

According to legend, ancestors of today's Oromo people in a region of Kaffa in Ethiopia were believed to have been the first to recognize the energizing effect of the coffee plant, though no direct evidence has been found indicating where in Africa coffee grew or who among the native populations might have used it as a stimulant or even known about it, earlier than the 17th century. The story of Kaldi, the 9th-century Ethiopian goatherd who discovered coffee when he noticed how excited his goats became after eating the beans from a coffee plant, did not appear in writing until 1671 and is probably apocryphal.

Other accounts attribute the discovery of coffee to Sheikh Omar. According to the ancient chronicle (preserved in the Abd-Al-Kadir manuscript), Omar, who was known for his ability to cure the sick through prayer, was once exiled from Mocha in Yemen to a desert cave near Ousab (modern day Wusab, about 90 km east of Zabid). Starving, Omar chewed berries from nearby shrubbery, but found them to be bitter. He tried roasting the seeds to improve the flavor, but they became hard. He then tried boiling them to soften the seed, which resulted in a fragrant brown liquid. Upon drinking the liquid Omar was revitalized and sustained for days. As stories of this "miracle drug" reached Mocha, Omar was asked to return and was made a saint. From Ethiopia, the coffee plant was introduced into the Arab World through Egypt and Yemen.

Historical Transmission

The earliest credible evidence of coffee-drinking or knowledge of the coffee tree appears in the middle of the 15th century in the accounts of Ahmed al-Ghaffar in Yemen. It was here in Arabia that coffee seeds were first roasted and brewed, in a similar way to how it is now prepared. Coffee was used by Sufi circles to stay awake for their religious rituals. Accounts differ on the origin of coffee (seeds) prior to its appearance in Yemen. One account credits Muhammad ben Said for bringing the beverage to Aden from the African coast. Other early accounts say Ali ben Omar of the Shadhili Sufi order was the first to introduce coffee to Arabia. According to al Shardi, Ali ben Omar may have encountered coffee during his stay with the Adal king Sadadin's companions in 1401. Famous 16th century Islamic scholar Ibn Hajar al-Haytami notes in his writings of a beverage called qahwa developed from a tree in the Zeila region.

A late 19th century advertisement for coffee essence

A coffee can from the first half of the 20th century. From the Museo del Objeto del Objeto collection.

By the 16th century, it had reached the rest of the Middle East, Persia, Turkey, and northern Africa. The first coffee smuggled out of the Middle East was by Sufi Baba Budan from Yemen to India in 1670. Before then, all exported coffee was boiled or otherwise sterilised. Portraits of Baba Budan depict him as having smuggled seven coffee seeds by strapping them to his chest. The first plants grown from these smuggled seeds were planted in Mysore. Coffee then spread to Italy, and to the rest of Europe, to Indonesia, and to the Americas.

In 1583, Leonhard Rauwolf, a German physician, gave this description of coffee after returning from a ten-year trip to the Near East:

A beverage as black as ink, useful against numerous illnesses, particularly those of the stomach. Its consumers take it in the morning, quite frankly, in a porcelain cup that is passed around and from which each one drinks a cupful. It is composed of water and the fruit from a bush called bunnu.

— Léonard Rauwolf, Reise in die Morgenländer (in German)

From the Middle East, coffee spread to Italy. The thriving trade between Venice and North Africa, Egypt, and the Middle East brought many goods, including coffee, to the Venetian port. From Venice, it was introduced to the rest of Europe. Coffee became more widely accepted after it was deemed a Christian beverage by Pope Clement VIII in 1600, despite appeals to ban the "Muslim drink." The first European coffee house opened in Rome in 1645.

A 1919 advertisement for *G Washington's Coffee*. The first instant coffee was invented by inventor George Washington in 1909.

The Dutch East India Company was the first to import coffee on a large scale. The Dutch later grew the crop in Java and Ceylon. The first exports of Indonesian coffee from Java to the Netherlands occurred in 1711.

Through the efforts of the British East India Company, coffee became popular in England as well. Oxford's Queen's Lane Coffee House, established in 1654, is still in existence today. Coffee was introduced in France in 1657, and in Austria and Poland after the 1683 Battle of Vienna, when coffee was captured from supplies of the defeated Turks.

When coffee reached North America during the Colonial period, it was initially not as successful as it had been in Europe as alcoholic beverages remained more popular. During the Revolutionary War, the demand for coffee increased so much that dealers had to hoard their scarce supplies and raise prices dramatically; this was also due to the reduced availability of tea from British merchants, and a general resolution among many Americans to avoid drinking tea following the 1773 Boston Tea Party.

After the War of 1812, during which Britain temporarily cut off access to tea imports, the Americans' taste for coffee grew. Coffee consumption declined in England, giving way to tea during the 18th century. The latter beverage was simpler to make, and had become cheaper with the British conquest of India and the tea industry there. During the Age of Sail, seamen aboard ships of the British Royal Navy made substitute coffee by dissolving burnt bread in hot water.

The Frenchman Gabriel de Clieu took a coffee plant to the French territory of Martinique in the Caribbean, from which much of the world's cultivated arabica coffee is descended. Coffee thrived in the climate and was conveyed across the Americas. Coffee was cultivated in Saint-Domingue (now Haiti) from 1734, and by 1788 it supplied half the world's coffee. The conditions that the slaves worked in on coffee plantations were a factor in the soon to follow Haitian Revolution. The coffee industry never fully recovered there. It made a brief come-back in 1949 when Haiti was the world's 3rd largest coffee exporter, but fell quickly into rapid decline.

Meanwhile, coffee had been introduced to Brazil in 1727, although its cultivation did not gather momentum until independence in 1822. After this time massive tracts of rainforest were cleared for coffee plantations, first in the vicinity of Rio de Janeiro and later São Paulo. Brazil went from having essentially no coffee exports in 1800, to being a significant regional producer in 1830, to being the largest producer in the world by 1852. In 1910-20, Brazil exported around 70% of the world's coffee, Colombia, Guatemala, and Venezuela, exported half of the remaining 30%, and Old World production accounted for less than 5% of world exports.

Cultivation was taken up by many countries in Central America in the latter half of the 19th century, and almost all involved the large-scale displacement and exploitation of the indigenous people. Harsh conditions led to many uprisings, coups and bloody suppression of peasants. The notable exception was Costa Rica, where lack of ready labor prevented the formation of large farms. Smaller farms and more egalitarian conditions ameliorated unrest over the 19th and 20th centuries.

Rapid growth in coffee production in South America during the second half of the 19th century was matched by growth in consumption in developed countries, though nowhere has this growth been as pronounced as in the United States, where high rate of population growth was compounded by doubling of per capita consumption between 1860 and 1920. Though the United States was not the heaviest coffee-drinking nation at the time (Nordic countries, Belgium, and Netherlands all had comparable or higher levels of per capita consumption), due to its sheer size, it was already the largest consumer of coffee in the world by 1860, and, by 1920, around half of all coffee produced worldwide was consumed in the USA.

Coffee has become a vital cash crop for many developing countries. Over one hundred million people in developing countries have become dependent on coffee as their primary source of income. It has become the primary export and backbone for African countries like Uganda, Burundi, Rwanda, and Ethiopia, as well as many Central American countries.

Biology

Several species of shrub of the genus *Coffea* produce the berries from which coffee is extracted. The two main species commercially cultivated are *Coffea canephora* (predominantly a form known as 'robusta') and *C. arabica*. *C. arabica*, the most highly regarded species, is native to the southwestern highlands of Ethiopia and the Boma Plateau in southeastern Sudan and possibly Mount Marsabit in northern Kenya. *C. canephora* is native to western and central Subsaharan Africa, from Guinea to Uganda and southern Sudan. Less popular species are *C. liberica, C. stenophylla, C. mauritiana*, and *C. racemosa.*

Illustration of *Coffea arabica* plant and seeds

All coffee plants are classified in the large family Rubiaceae. They are evergreen shrubs or trees that may grow 5 m (15 ft) tall when unpruned. The leaves are dark green and glossy, usually 10–15 cm (4–6 in) long and 6 cm (2.4 in) wide, simple, entire, and opposite. Petioles of opposite leaves fuse at base to form interpetiolar stipules, characteristic of Rubiaceae. The flowers are axillary, and clusters of fragrant white flowers bloom simultaneously. Gynoecium consists of inferior ovary, also characteristic of Rubiaceae. The flowers are followed by oval berries of about 1.5 cm (0.6 in). When immature they are green, and they ripen to yellow, then crimson, before turning black on drying. Each berry usually contains two seeds, but 5–10% of the berries have only one; these are called peaberries. Arabica berries ripen in six to eight months, while robusta take nine to eleven months.

Robusta coffee flowers

Coffea arabica is predominantly self-pollinating, and as a result the seedlings are generally uniform and vary little from their parents. In contrast, *Coffea canephora*, and *C. liberica* are self-in-compatible and require outcrossing. This means that useful forms and hybrids must be propagated vegetatively. Cuttings, grafting, and budding are the usual methods of vegetative propagation. On the other hand, there is great scope for experimentation in search of potential new strains.

In 2016, Oregon State University entomologist George Poinar, Jr. announced the discovery of a new plant species that's a 45-million-year-old relative of coffee found in amber. Named Strychnos electri, after the Greek word for amber (electron), the flowers represent the first-ever fossils of an asterid, which is a family of flowering plants that not only later gave us coffee, but also sunflowers, peppers, potatoes, mint — and deadly poisons.

Cultivation

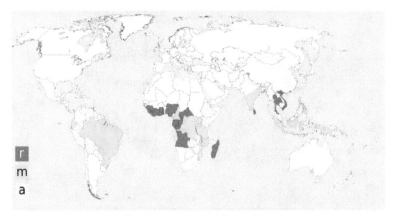

Map showing areas of coffee cultivation:
r:*Coffea canephora*
m:*Coffea canephora* and *Coffea arabica*
a:*Coffea arabica*

The traditional method of planting coffee is to place 20 seeds in each hole at the beginning of the rainy season. This method loses about 50% of the seeds' potential, as about half fail to sprout. A more effective method of growing coffee, used in Brazil, is to raise seedlings in nurseries that are then planted outside at six to twelve months. Coffee is often intercropped with food crops, such as corn, beans, or rice during the first few years of cultivation as farmers become familiar with its requirements. Coffee plants grow within a defined area between the tropics of Cancer and Capricorn, termed the bean belt or coffee belt.

Of the two main species grown, arabica coffee (from *C. arabica*) is generally more highly regarded than robusta coffee (from *C. canephora*); robusta tends to be bitter and have less flavor but better body than arabica. For these reasons, about three-quarters of coffee cultivated worldwide is *C. arabica*. Robusta strains also contain about 40–50% more caffeine than arabica. Consequently, this species is used as an inexpensive substitute for arabica in many commercial coffee blends. Good quality robusta beans are used in traditional Italian espresso blends to provide a full-bodied taste and a better foam head (known as *crema*).

Additionally, *Coffea canephora* is less susceptible to disease than *C. arabica* and can be cultivated in lower altitudes and warmer climates where *C. arabica* will not thrive. The robusta strain was first collected in 1890 from the Lomani River, a tributary of the Congo River, and was conveyed from Zaire (now the Democratic Republic of Congo) to Brussels to Java around 1900. From Java, further breeding resulted in the establishment of robusta plantations in many countries. In particular, the spread of the devastating coffee leaf rust (*Hemileia vastatrix*), to which *C. arabica* is vulnerable, hastened the uptake of the resistant robusta. Coffee leaf rust is found in virtually all countries that produce coffee.

Over 900 species of insect have been recorded as pests of coffee crops worldwide. Of these, over a third are beetles, and over a quarter are bugs. Some 20 species of nematodes, 9 species of mites, and several snails and slugs also attack the crop. Birds and rodents sometimes eat coffee berries, but their impact is minor compared to invertebrates. In general, *arabica* is the more sensitive species to invertebrate predation overall. Each part of the coffee plant is assailed by different animals. Nematodes attack the roots, coffee borer beetles burrow into stems and woody material, and the foliage is attacked by over 100 species of larvae (caterpillars) of butterflies and moths.

Mass spraying of insecticides has often proven disastrous, as predators of the pests are more sensitive than the pests themselves. Instead, integrated pest management has developed, using techniques such as targeted treatment of pest outbreaks, and managing crop environment away from conditions favouring pests. Branches infested with scale are often cut and left on the ground, which promotes scale parasites to not only attack the scale on the fallen branches but in the plant as well.

The 2-mm-long coffee borer beetle (*Hypothenemus hampei*) is the most damaging insect pest to the world's coffee industry, destroying up to 50 percent or more of the coffee berries on plantations in most coffee-producing countries. The adult female beetle nibbles a single tiny hole in a coffee berry and lays 35 to 50 eggs. Inside, the offspring grow, mate, and then emerge from the commercially ruined berry to disperse, repeating the cycle. Pesticides are mostly ineffective because the beetle juveniles are protected inside the berry nurseries, but they are vulnerable to predation by birds when they emerge. When groves of trees are nearby, the American yellow warbler, rufous-capped warbler, and other insectivorous birds have been shown to reduce by 50 percent the number of coffee berry borers in Costa Rica coffee plantations.

Beans from different countries or regions can usually be distinguished by differences in flavor, aroma, body, and acidity. These taste characteristics are dependent not only on the coffee's growing region, but also on genetic subspecies (varietals) and processing. Varietals are generally known by the region in which they are grown, such as Colombian, Java and Kona.

Arabica coffee beans are cultivated mainly in Latin America, eastern Africa or Asia, while robusta beans are grown in central Africa, throughout southeast Asia, and Brazil.

Ecological Effects

Originally, coffee farming was done in the shade of trees that provided a habitat for many animals and insects. Remnant forest trees were used for this purpose, but many species have been planted as well. These include leguminous trees of the genera *Acacia, Albizia, Cassia, Erythrina, Gliricidia, Inga,* and *Leucaena,* as well as the nitrogen-fixing non-legume sheoaks of the genus *Casuarina,* and the silky oak *Grevillea robusta.*

This method is commonly referred to as the traditional shaded method, or "shade-grown". Starting in the 1970s, many farmers switched their production method to sun cultivation, in which coffee is grown in rows under full sun with little or no forest canopy. This causes berries to ripen more rapidly and bushes to produce higher yields, but requires the clearing of trees and increased use of fertilizer and pesticides, which damage the environment and cause health problems.

A flowering *Coffea arabica* tree in a Brazilian plantation

Unshaded coffee plants grown with fertilizer yield the most coffee, although unfertilized shaded crops generally yield more than unfertilized unshaded crops: the response to fertilizer is much greater in full sun. While traditional coffee production causes berries to ripen more slowly and produce lower yields, the quality of the coffee is allegedly superior. In addition, the traditional shaded method provides living space for many wildlife species. Proponents of shade cultivation say environmental problems such as deforestation, pesticide pollution, habitat destruction, and soil and water degradation are the side effects of the practices employed in sun cultivation.

The American Birding Association, Smithsonian Migratory Bird Center, National Arbor Day Foundation, and the Rainforest Alliance have led a campaign for 'shade-grown' and organic coffees, which can be sustainably harvested. Shaded coffee cultivation systems show greater biodiversity than full-sun systems, and those more distant from continuous forest compare rather poorly to undisturbed native forest in terms of habitat value for some bird species.

Another issue concerning coffee is its use of water. It takes about 140 liters (37 U.S. gal) of water to grow the coffee beans needed to produce one cup of coffee, and coffee is often grown in countries where there is a water shortage, such as Ethiopia.

Used coffee grounds may be used for composting or as a mulch. They are especially appreciated by worms and acid-loving plants such as blueberries. Some commercial coffee shops run initiatives to make better use of these grounds, including Starbucks' "Grounds for your Garden" project, and community sponsored initiatives such as "Ground to Ground".

Starbucks sustainability chief Jim Hanna has warned that climate change may significantly impact coffee yields within a few decades. A study by Kew Royal Botanic Gardens concluded that global warming threatens the genetic diversity of Arabica plants found in Ethiopia and surrounding countries.

Production

Top five green coffee producers in 2013		
Rank	**Country**	**Millions of tonnes**
1	Brazil	3.0
2	Vietnam	1.5
3	Indonesia	0.7
4	Colombia	0.7
5	India	0.3
	World	**8.9**

In 2013, world production of green coffee was 8.9 million tonnes, with Brazil as the leader in production followed by Vietnam, Indonesia, Colombia and India (table).

Processing

Traditional coffee beans drying in Kalibaru, Indonesia

Coffee berries and their seeds undergo several processes before they become the familiar roasted coffee. Berries have been traditionally selectively picked by hand; a labor-intensive method, it involves the selection of only the berries at the peak of ripeness. More commonly, crops are strip picked, where all berries are harvested simultaneously regardless of ripeness by person or machine. After picking, green coffee is processed by one of two methods—the dry process method, simpler and less labor-intensive as the berries can be strip picked, and the wet process method, which incorporates fermentation into the process and yields a mild coffee.

Then they are sorted by ripeness and color and most often the flesh of the berry is removed, usually by machine, and the seeds are fermented to remove the slimy layer of mucilage still present on the seed. When the fermentation is finished, the seeds are washed with large quantities of fresh water to remove the fermentation residue, which generates massive amounts of coffee wastewater. Finally, the seeds are dried.

The best (but least used) method of drying coffee is using drying tables. In this method, the pulped and fermented coffee is spread thinly on raised beds, which allows the air to pass on all sides of the coffee, and then the coffee is mixed by hand. In this method the drying that takes place is more uniform, and fermentation is less likely. Most African coffee is dried in this manner and certain coffee farms around the world are starting to use this traditional method.

Next, the coffee is sorted, and labeled as green coffee. Another way to let the coffee seeds dry is to let them sit on a concrete patio and rake over them in the sunlight. Some companies use cylinders to pump in heated air to dry the coffee seeds, though this is generally in places where the humidity is very high.

An Asian coffee known as kopi luwak undergoes a peculiar process made from coffee berries eaten by the Asian palm civet, passing through its digestive tract, with the beans eventually harvested from feces. Coffee brewed from this process is among the most expensive in the world, with bean prices reaching $160 per pound or $30 per brewed cup. Kopi luwak coffee is said to have uniquely rich, slightly smoky aroma and flavor with hints of chocolate, resulting from the action of digestive enzymes breaking down bean proteins to facilitate partial fermentation.

Roasting

The next step in the process is the roasting of the green coffee. Coffee is usually sold in a roasted state, and with rare exceptions all coffee is roasted before it is consumed. It can be sold roasted by the supplier, or it can be home roasted. The roasting process influences the taste of the beverage by changing the coffee bean both physically and chemically. The bean decreases in weight as moisture is lost and increases in volume, causing it to become less dense. The density of the bean also influences the strength of the coffee and requirements for packaging.

The actual roasting begins when the temperature inside the bean reaches approximately 200 °C (392 °F), though different varieties of seeds differ in moisture and density and therefore roast at different rates. During roasting, caramelization occurs as intense heat breaks down starches, changing them to simple sugars that begin to brown, which alters the color of the bean.

Sucrose is rapidly lost during the roasting process, and may disappear entirely in darker roasts. During roasting, aromatic oils and acids weaken, changing the flavor; at 205 °C (401 °F), other oils start to develop. One of these oils, caffeol, is created at about 200 °C (392 °F), which is largely responsible for coffee's aroma and flavor.

Roasting is the last step of processing the beans in their intact state. During this last treatment, while still in the bean state, more caffeine breaks down above 235 °C (455 °F). Dark roasting is the utmost step in bean processing removing the most caffeine. Although, dark roasting is not to be confused with the Decaffeination process.

Grading Roasted Beans

Coffee "cuppers", or professional tasters, grade the coffee

Depending on the color of the roasted beans as perceived by the human eye, they will be labeled as light, medium light, medium, medium dark, dark, or very dark. A more accurate method of discerning the degree of roast involves measuring the reflected light from roasted seeds illuminated with a light source in the near-infrared spectrum. This elaborate light meter uses a process known as spectroscopy to return a number that consistently indicates the roasted coffee's relative degree of roast or flavor development.

Roast Characteristics

The degree of roast has an effect upon coffee flavor and body. Darker roasts are generally bolder because they have less fiber content and a more sugary flavor. Lighter roasts have a more complex and therefore perceived stronger flavor from aromatic oils and acids otherwise destroyed by longer roasting times. Roasting does not alter the amount of caffeine in the bean, but does give less caffeine when the beans are measured by volume because the beans expand during roasting.

A small amount of chaff is produced during roasting from the skin left on the seed after processing. Chaff is usually removed from the seeds by air movement, though a small amount is added to dark roast coffees to soak up oils on the seeds.

Decaffeination

Decaffeination may also be part of the processing that coffee seeds undergo. Seeds are decaffeinated when they are still green. Many methods can remove caffeine from coffee, but all involve either soaking the green seeds in hot water (often called the "Swiss water process") or steaming them, then using a solvent to dissolve caffeine-containing oils. Decaffeination is often done by processing companies, and the extracted caffeine is usually sold to the pharmaceutical industry.

Storage

Coffee is best stored in an airtight container made of ceramic, glass, or non-reactive metal. Higher quality prepackaged coffee usually has a one-way valve which prevents air from entering while allowing the coffee to release gases. Coffee freshness and flavor is preserved when it is stored away from moisture, heat, and light. The ability of coffee to absorb strong smells from food means that it should be kept away from such smells. Storage of coffee in the refrigerator is not recommended due to the presence of moisture which can cause deterioration. Exterior walls of buildings which face the sun may heat the interior of a home, and this heat may damage coffee stored near such a wall. Heat from nearby ovens also harms stored coffee.

In 1931, a method of packing coffee in a sealed vacuum in cans was introduced. The roasted coffee was packed and then 99% of the air was removed, allowing the coffee to be stored indefinitely until the can was opened. Today this method is in mass use for coffee in a large part of the world.

Brewing

Coffee beans must be ground and brewed to create a beverage. The criteria for choosing a method include flavor and economy. Almost all methods of preparing coffee require that the beans be ground and then mixed with hot water long enough to allow the flavor to emerge but not so long as to draw out bitter compounds. The liquid can be consumed after the spent grounds are removed. Brewing considerations include the fineness of grind, the way in which the water is used to extract the flavor, the ratio of coffee grounds to water (the brew ratio), additional flavorings such as sugar, milk, and spices, and the technique to be used to separate spent grounds. Ideal holding temperatures range from 85–88 °C (185–190 °F) to as high as 93 °C (199 °F) and the ideal serving temperature is 68 to 79 °C (154 to 174 °F). The recommended brew ratio for non-espresso coffee is around 55 to 60 grams of grounds per litre of water, or two level tablespoons for a 5- or 6-ounce cup.

The roasted coffee beans may be ground at a roastery, in a grocery store, or in the home. Most coffee is roasted and ground at a roastery and sold in packaged form, though roasted coffee beans can be ground at home immediately before consumption. It is also possible, though uncommon, to roast raw beans at home.

The choice of brewing method depends to some extent on the degree to which the coffee beans have been roasted. Lighter roasted coffee tends to be used for filter coffee as the combination of method and roast style results in higher acidity, complexity, and clearer nuances. Darker roasted coffee is used for espresso because the machine naturally extracts more dissolved solids, causing lighter coffee to become too acidic.

Coffee beans may be ground in several ways. A burr grinder uses revolving elements to shear the seed; a blade grinder cuts the seeds with blades moving at high speed; and a mortar and pestle crushes the seeds. For most brewing methods a burr grinder is deemed superior because the grind is more even and the grind size can be adjusted.

The type of grind is often named after the brewing method for which it is generally used. Turkish grind is the finest grind, while coffee percolator or French press are the coarsest grinds. The most common grinds are between these two extremes: a medium grind is used in most home coffee-brewing machines.

Coffee may be brewed by several methods. It may be boiled, steeped, or pressurized. Brewing coffee by boiling was the earliest method, and Turkish coffee is an example of this method. It is prepared by grinding or pounding the seeds to a fine powder, then adding it to water and bringing it to the boil for no more than an instant in a pot called a *cezve* or, in Greek, a *bríki*. This produces a strong coffee with a layer of foam on the surface and sediment (which is not meant for drinking) settling at the bottom of the cup.

Coffee percolators and automatic coffeemakers brew coffee using gravity. In an automatic coffee-maker, hot water drips onto coffee grounds that are held in a paper, plastic, or perforated metal coffee filter, allowing the water to seep through the ground coffee while extracting its oils and essences. The liquid drips through the coffee and the filter into a carafe or pot, and the spent grounds are retained in the filter.

In a percolator, boiling water is forced into a chamber above a filter by steam pressure created by boiling. The water then seeps through the grounds, and the process is repeated until terminated by removing from the heat, by an internal timer, or by a thermostat that turns off the heater when the entire pot reaches a certain temperature.

Coffee may be brewed by steeping in a device such as a French press (also known as a cafetière, coffee press or coffee plunger). Ground coffee and hot water are combined in a cylindrical vessel and left to brew for a few minutes. A circular filter which fits tightly in the cylinder fixed to a plunger is then pushed down from the top to force the grounds to the bottom. The filter retains the grounds at the bottom as the coffee is poured from the container. Because the coffee grounds are in direct contact with the water, all the coffee oils remain in the liquid, making it a stronger beverage. This method of brewing leaves more sediment than in coffee made by an automatic coffee machine. Supporters of the French press method point out that the sediment issue can be minimized by using the right type of grinder: they claim that a rotary blade grinder cuts the coffee bean into a wide range of sizes, including a fine coffee dust that remains as sludge at the bottom of the cup, while a burr grinder uniformly grinds the beans into consistently-sized grinds, allowing the coffee to settle uniformly and be trapped by the press. Within the first minute of brewing 95% of the caffeine is released from the coffee bean.

The espresso method forces hot pressurized and vaporized water through ground coffee. As a result of brewing under high pressure (ideally between 9–10 atm), the espresso beverage is more concentrated (as much as 10 to 15 times the quantity of coffee to water as gravity-brewing methods can produce) and has a more complex physical and chemical constitution. A well-prepared espresso has a reddish-brown foam called *crema* that floats on the surface. Other pressurized water methods include the moka pot and vacuum coffee maker.

Cold brew coffee is made by steeping coarsely ground beans in cold water for several hours, then filtering them. This results in a brew lower in acidity than most hot-brewing methods.

Nutrition

Brewed coffee from typical grounds prepared with tap water contains 40 mg caffeine per 100 gram and no essential nutrients in significant content. In espresso, however, likely due to its higher amount of suspended solids, there are significant contents of magnesium, the B vitamins, niacin and riboflavin, and 212 mg of caffeine per 100 grams of grounds.

Serving

Once brewed, coffee may be served in a variety of ways. Drip-brewed, percolated, or French-pressed/cafetière coffee may be served as *white coffee* with a dairy product such as milk or cream, or dairy substitute, or as *black coffee* with no such addition. It may be sweetened with sugar or artificial sweetener. When served cold, it is called *iced coffee*.

Espresso-based coffee has a variety of possible presentations. In its most basic form, an espresso is served alone as a *shot* or *short black*, or with hot water added, when it is known as Caffè Americano. A long black is made by pouring a double espresso into an equal portion of water, retaining the crema, unlike Caffè Americano. Milk is added in various forms to an espresso: steamed milk makes a caffè latte, equal parts steamed milk and milk froth make a cappuccino, and a dollop of hot foamed milk on top creates a caffè macchiato. A flat white is prepared by adding steamed hot milk (microfoam) to espresso so that the flavour is brought out and the texture is unusually velvety. It has less milk than a latte but both are varieties of coffee to which the milk can be added in such a way as to create a decorative surface pattern. Such effects are known as latte art.

Coffee can also be incorporated with alcohol to produce a variety of beverages: it is combined with whiskey in Irish coffee, and it forms the base of alcoholic coffee liqueurs such as Kahlúa and Tia Maria. Darker beers such as stout and porter give a chocolate or coffee-like taste due to roasted grains even though actual coffee beans are not added to it.

Instant Coffee

A number of products are sold for the convenience of consumers who do not want to prepare their own coffee. Instant coffee is dried into soluble powder or freeze-dried into granules that can be quickly dissolved in hot water. Originally invented in 1907, it rapidly gained in popularity in many countries in the post-war period, with Nescafé being the most popular product. Many consumers determined that the convenience in preparing a cup of instant coffee more than made up for a perceived inferior taste, although, since the late 1970s, instant coffee has been produced differently in such a way that is similar to the taste of freshly brewed coffee. Paralleling (and complementing) the rapid rise of instant coffee was the coffee vending machine invented in 1947 and widely distributed since the 1950s.

Canned coffee has been popular in Asian countries for many years, particularly in China, Japan, South Korea, and Taiwan. Vending machines typically sell varieties of flavored canned coffee, much like brewed or percolated coffee, available both hot and cold. Japanese convenience stores and groceries also have a wide availability of bottled coffee drinks, which are typically lightly sweetened and pre-blended with milk. Bottled coffee drinks are also consumed in the United States.

Liquid coffee concentrates are sometimes used in large institutional situations where coffee needs to be produced for thousands of people at the same time. It is described as having a flavor about as good as low-grade robusta coffee, and costs about 10¢ a cup to produce. The machines can process up to 500 cups an hour, or 1,000 if the water is preheated.

Sale and Distribution

Coffee ingestion on average is about a third of that of tap water in North America and Europe.

Worldwide, 6.7 million metric tons of coffee were produced annually in 1998–2000, and the forecast is a rise to seven million metric tons annually by 2010.

Brazilian coffee sacks

Brazil remains the largest coffee exporting nation, however Vietnam tripled its exports between 1995 and 1999 and became a major producer of robusta seeds. Indonesia is the third-largest coffee exporter overall and the largest producer of washed arabica coffee. Organic Honduran coffee is a rapidly growing emerging commodity owing to the Honduran climate and rich soil.

In 2013, *The Seattle Times* reported that global coffee prices dropped more than 50 percent year-over-year. In Thailand, black ivory coffee beans are fed to elephants whose digestive enzymes reduce the bitter taste of beans collected from dung. These beans sell for up to $1,100 a kilogram ($500 per lb), achieving the world's most expensive coffee some three times costlier than beans harvested from the dung of Asian palm civets.

Commodity Market

Coffee is bought and sold as green coffee beans by roasters, investors, and price speculators as a tradable commodity in commodity markets and exchange-traded funds. Coffee futures contracts for Grade 3 washed arabicas are traded on the New York Mercantile Exchange under ticker symbol KC, with contract deliveries occurring every year in March, May, July, September, and December. Coffee is an example of a product that has been susceptible to significant commodity futures price variations. Higher and lower grade arabica coffees are sold through other channels. Futures contracts for robusta coffee are traded on the London International Financial Futures and Options Exchange and, since 2007, on the New York Intercontinental Exchange.

Dating to the 1970s, coffee has been incorrectly described by many, including historian Mark Pendergrast, as the world's "second most legally traded commodity". Instead, "coffee was the second most valuable commodity exported by developing countries," from 1970 to circa 2000. This fact was derived from the United Nations Conference on Trade and Development Commodity Yearbooks which show "Third World" commodity exports by value in the period 1970–1998 as being in order of crude oil in first place, coffee in second, followed by sugar, cotton, and others. Coffee continues to be an important commodity export for developing countries, but more recent figures are not readily available due to the shifting and politicized nature of the category "developing country".

International Coffee Day, which is claimed to have originated in Japan in 1983 with an event organised by the All Japan Coffee Association, takes place on September 29 in several countries.

Health and Pharmacology

Method of Action

Skeletal structure of a caffeine molecule

The primary psychoactive chemical in coffee is caffeine, an adenosine antagonist that is known for its stimulant effects. Coffee also contains the monoamine oxidase inhibitors β-carboline and harmane, which may contribute to its psychoactivity.

In a healthy liver, caffeine is mostly broken down by the hepatic microsomal enzymatic system. The excreted metabolites are mostly paraxanthines—theobromine and theophylline—and a small amount of unchanged caffeine. Therefore, the metabolism of caffeine depends on the state of this enzymatic system of the liver.

Polyphenols in coffee have been shown to affect free radicals in vitro, but there is no evidence that this effect occurs in humans. Polyphenol levels vary depending on how beans are roasted as well as for how long. As interpreted by the Linus Pauling Institute and the European Food Safety Authority, dietary polyphenols, such as those ingested by consuming coffee, have little or no direct antioxidant value following ingestion.

Health Effects

Findings have been contradictory as to whether coffee has any specific health benefits, and results are similarly conflicting regarding the potentially harmful effects of coffee consumption. Furthermore, results and generalizations are complicated by differences in age, gender, health status, and serving size.

Extensive scientific research has been conducted to examine the relationship between coffee consumption and an array of medical conditions. The consensus in the medical community is that moderate regular coffee drinking in healthy individuals is either essentially benign or mildly beneficial. Researchers involved in an ongoing 22-year study by the Harvard School of Public Health stated that "Coffee may have potential health benefits, but more research needs to be done."

Mortality

In 2012, the National Institutes of Health–AARP Diet and Health Study analysed the relationship between coffee drinking and mortality. They found that higher coffee consumption was associated with lower risk of death, and that those who drank any coffee lived longer than those who did not. However the authors noted, "whether this was a causal or associational finding cannot be determined from our data." A 2014 meta-analysis found that coffee consumption (4 cups/day) was inversely associated with all-cause mortality (a 16% lower risk), as well as cardiovascular disease mortality specifically (a 21% lower risk from drinking 3 cups/day), but not with cancer mortality. Additional meta-analysis studies corroborated these findings, showing that higher coffee consumption (2–4 cups per day) was associated with a reduced risk of death by all disease causes.

Cardiovascular Disease

Coffee is no longer thought to be a risk factor for coronary heart disease. A 2012 meta-analysis concluded that people who drank moderate amounts of coffee had a lower rate of heart failure, with the biggest effect found for those who drank more than four cups a day. Moreover, in one preliminary study, habitual coffee consumption was associated with improved vascular function. Interestingly, a recent meta-analysis showed that coffee consumption was associated with a reduced risk of death in patients who have had a myocardial infarction.

Mental Health

One review published in 2004 indicated a negative correlation between suicide rates and coffee consumption, but this effect has not been confirmed in larger studies.

Long-term studies of both risk and potential benefit of coffee consumption by elderly people, including assessment on symptoms of Alzheimer's disease and cognitive impairment, are not conclusive.

Some research suggests that a minority of moderate regular caffeine consumers experience some amount of clinical depression, anxiety, low vigor, or fatigue when discontinuing their caffeine use. However, the methodology of these studies has been criticized. Withdrawal effects are more common and better documented in heavy caffeine users.

Coffee caffeine may aggravate pre-existing conditions such as migraines, arrhythmias, and cause sleep disturbances. Caffeine withdrawal from chronic use causes consistent effects typical of physical dependence, including headaches, mood changes and the possibility of reduced cerebral blood flow.

Type II Diabetes

In a systematic review and meta-analysis of 28 prospective observational studies, representing 1,109,272 participants, every additional cup of caffeinated and decaffeinated coffee consumed in a day was associated with a 9% (95% CI 6%, 11%) and 6% (95% CI 2%, 9%) lower risk of type 2 diabetes, respectively.

Cancer

The effects of coffee consumption on cancer risk remain unclear, with reviews and meta-analyses showing either no relationship or a small lower risk of cancer onset.

Risks

Instant coffee has a greater amount of acrylamide than brewed coffee. It was once thought that coffee aggravates gastroesophageal reflux disease but recent research suggests no link.

Caffeine Content

Depending on the type of coffee and method of preparation, the caffeine content of a single serving can vary greatly. The caffeine content of a cup of coffee varies depending mainly on the brewing method, and also on the variety of seed. According to the USDA National Nutrient Database, an 8-ounce (237 ml) cup of "coffee brewed from grounds" contains 95 mg caffeine, whereas an espresso (25 ml) contains 53 mg.

According to an article in the *Journal of the American Dietetic Association*, coffee has the following caffeine content, depending on how it is prepared:

	Serving size	Caffeine content
Brewed	7 oz, 207 ml	80–135 mg
Drip	7 oz, 207 ml	115–175 mg
Espresso	1.5–2 oz, 45–60 ml	100 mg

While the percent of caffeine content in coffee seeds themselves diminishes with increased roast level, the opposite is true for coffee brewed from different grinds and brewing methods using the same proportion of coffee to water volume. The coffee sack (similar to the French press and other steeping methods) extracts more caffeine from dark roasted seeds; the percolator and espresso methods extract more caffeine from light roasted seeds:[clarification needed What are the units?]

	Light roast	Medium roast	Dark roast
Coffee sack – coarse grind	0.046	0.045	0.054
Percolator – coarse grind	0.068	0.065	0.060
Espresso – fine grind	0.069	0.062	0.061

Coffea arabica normally contains about half the caffeine of *Coffea robusta*. A *Coffea arabica* bean containing very little caffeine was discovered in Ethiopia in 2004.

Coffeehouses

Widely known as coffeehouses or cafés, establishments serving prepared coffee or other hot beverages have existed for over five hundred years. Various legends involving the introduction of coffee to Istanbul at a "Kiva Han" in the late 15th century circulate in culinary tradition, but with no documentation.

Coffeehouse in Palestine (c.1900)

Coffeehouses in Mecca became a concern as places for political gatherings to the imams who banned them, and the drink, for Muslims between 1512 and 1524. In 1530 the first coffeehouse was opened in Damascus. The first coffeehouse in Constantinople was opened in 1475 by traders arriving from Damascus and Aleppo. Soon after, coffeehouses became part of the Ottoman Culture, spreading rapidly to all regions of the Ottoman Empire.

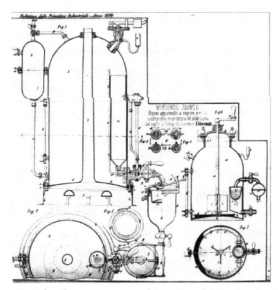

First patent for the espresso machine, Angelo Moriondo (1884)

In the 17th century, coffee appeared for the first time in Europe outside the Ottoman Empire, and coffeehouses were established and quickly became popular. The first coffeehouses in Western Europe appeared in Venice, as a result of the traffic between La Serenissima and the Ottomans; the very first one is recorded in 1645. The first coffeehouse in England was set up in Oxford in 1650 by a Jewish man named Jacob in the building now known as "The Grand Cafe". A plaque on the wall still commemorates this and the Cafe is now a trendy cocktail bar. By 1675, there were more than 3,000 coffeehouses in England.

A legend says that after the second Turkish siege of Vienna in 1683, the Viennese discovered many bags of coffee in the abandoned Ottoman encampment. Using this captured stock, a Polish soldier named Kulczycki opened the first coffeehouse in Vienna. This story never happened. Nowadays it is proven that the first coffeehouse in Vienna was opened by the Armenian Johannes Theodat in 1685.

In 1672 an Armenian named Pascal established a coffee stall in Paris that was ultimately unsuccessful and the city had to wait until 1689 for its first coffeehouse when Procopio Cutò opened the Café Procope. This coffeehouse still exists today and was a major meeting place of the French Enlightenment; Voltaire, Rousseau, and Denis Diderot frequented it, and it is arguably the birthplace of the *Encyclopédie*, the first modern encyclopedia. America had its first coffeehouse in Boston, in 1676. Coffee, tea and beer were often served together in establishments which functioned both as coffeehouses and taverns; one such was the Green Dragon in Boston, where John Adams, James Otis, and Paul Revere planned rebellion.

The modern espresso machine was invented in Milan in 1945 by Achille Gaggia, and from there spread across coffeehouses and restaurants across Italy and the rest of Europe and North America in the early 1950s. An Italian named Pino Riservato opened the first espresso bar, the Moka Bar, in Soho in 1952, and there were 400 such bars in London alone by 1956. Cappucino was particularly popular among English drinkers. Similarly in the United States, the espresso craze spread. North Beach in San Francisco saw the opening of the Caffe Trieste in 1957, which saw Beat Generation poets such as Allen Ginsberg and Bob Kaufman alongside bemused Italian immigrants. Similar such cafes existed in Greenwich Village and elsewhere.

The first Peet's Coffee & Tea store opened in 1966 in Berkeley, California by Dutch native Alfred Peet. He chose to focus on roasting batches with fresher, higher quality seeds than was the norm at the time. He was a trainer and supplier to the founders of Starbuck's.

The international coffeehouse chain Starbucks began as a modest business roasting and selling coffee beans in 1971, by three college students Jerry Baldwin, Gordon Bowker, and Zev Siegl. The first store opened on March 30, 1971 at the Pike Place Market in Seattle, followed by a second and third over the next two years. Entrepreneur Howard Schultz joined the company in 1982 as Director of Retail Operations and Marketing, and pushed to sell premade espresso coffee. The others were reluctant, but Schultz opened Il Giornale in Seattle in April 1986. He bought the other owners out in March 1987 and pushed on with plans to expand—from 1987 to the end of 1991, the chain (rebranded from Il Giornale to Starbucks) expanded to over 100 outlets. The company has 16,600 stores in over 40 countries worldwide.

Barista at work

South Korea experienced almost 900 percent growth in the number of coffee shops in the country between 2006 and 2011. The capital city Seoul now has the highest concentration of coffee shops in the world, with more than 10,000 cafes and coffeehouses.

A contemporary term for a person who makes coffee beverages, often a coffeehouse employee, is a barista. The Specialty Coffee Association of Europe and the Specialty Coffee Association of America have been influential in setting standards and providing training.

Social and Culture

A coffee-house in Constantinople (modern day Istanbul), 1826

Coffee is often consumed alongside (or instead of) breakfast by many at home or when eating out at diners or cafeterias. It is often served at the end of a formal meal, normally with a dessert, and at times with an after-dinner mint, especially when consumed at a restaurant or dinner party.

Coffee break area

Break

A coffee break is a routine social gathering for a snack, the consumption of a hot beverage such as coffee or tea and short downtime practiced by employees in business and industry, corresponding with the Commonwealth terms "elevenses", "Smoko" (in Australia), "morning tea", "tea break", or even just "tea". An afternoon coffee break, or afternoon tea, sometimes occurs as well.

The coffee break originated in the late 19th century in Stoughton, Wisconsin, with the wives of Norwegian immigrants. The city celebrates this every year with the Stoughton Coffee Break Festival. In 1951, *Time* noted that "[s]ince the war, the coffee break has been written into union contracts". The term subsequently became popular through a Pan-American Coffee Bureau ad campaign of 1952 which urged consumers, "Give yourself a Coffee-Break – and Get What Coffee Gives to You." John B. Watson, a behavioral psychologist who worked with Maxwell House later in his career, helped to popularize coffee breaks within the American culture. Coffee breaks usually last from 10 to 20 minutes and frequently occur at the end of the first third of the work shift. In some companies and some civil service, the coffee break may be observed formally at a set hour. In some places, a "cart" with hot and cold beverages and cakes, breads and pastries arrives at the same time morning and afternoon, an employer may contract with an outside caterer for daily service, or coffee breaks may take place away from the actual work-area in a designated cafeteria or tea room. More generally, the phrase "coffee break" has also come to denote any break from work.

Prohibition

The Coffee Bearer, Orientalist painting by John Frederick Lewis (1857)

Coffee was initially used for spiritual reasons. At least 1,100 years ago, traders brought coffee across the Red Sea into Arabia (modern-day Yemen), where Muslim dervishes began cultivating the shrub in their gardens. At first, the Arabians made wine from the pulp of the fermented coffee berries. This beverage was known as *qishr* (*kisher* in modern usage) and was used during religious ceremonies.

Coffee drinking was prohibited by jurists and scholars (*ulema*) meeting in Mecca in 1511 as *haraam*, but the subject of whether it was intoxicating was hotly debated over the next 30 years until the ban was finally overturned in the mid-16th century. Use in religious rites among the Sufi branch of Islam led to coffee's being put on trial in Mecca: it was accused of being a heretical substance, and its production and consumption were briefly repressed. It was later prohibited in Ottoman Turkey under an edict by the Sultan Murad IV.

Coffee, regarded as a Muslim drink, was prohibited by Ethiopian Orthodox Christians until as late as 1889; it is now considered a national drink of Ethiopia for people of all faiths. Its early association in Europe with rebellious political activities led to Charles II outlawing coffeehouses from January 1676 (although the uproar created forced the monarch to back down two days before the ban was due to come into force). Frederick the Great banned it in Prussia in 1777 for nationalistic and economic reasons; concerned about the price of import, he sought to force the public back to consuming beer. Lacking coffee-producing colonies, Prussia had to import all its coffee at a great cost.

A contemporary example of religious prohibition of coffee can be found in The Church of Jesus Christ of Latter-day Saints. The organization holds that it is both physically and spiritually unhealthy to consume coffee. This comes from the Mormon doctrine of health, given in 1833 by founder Joseph Smith in a revelation called the Word of Wisdom. It does not identify coffee by name, but includes the statement that "hot drinks are not for the belly," which has been interpreted to forbid both coffee and tea.

Quite a number of members of the Seventh-day Adventist Church also avoid caffeinated drinks. In its teachings, the Church encourages members to avoid tea, coffee, and other stimulants. Abstinence from coffee, tobacco, and alcohol by many Adventists has afforded a near-unique opportunity for studies to be conducted within that population group on the health effects of coffee drinking, free from confounding factors. One study was able to show a weak but statistically significant association between coffee consumption and mortality from ischemic heart disease, other cardiovascular disease, all cardiovascular diseases combined, and all causes of death.

For a time, there had been controversy in the Jewish community over whether the coffee seed was a legume and therefore prohibited for Passover. Upon petition from coffeemaker Maxwell House, the coffee seed was classified in 1923 as a berry rather than a seed by orthodox Jewish rabbi Hersch Kohn, and therefore kosher for Passover.

Fair Trade

Small-sized bag of coffee beans

The concept of fair trade labeling, which guarantees coffee growers a negotiated preharvest price, began in the late 1980s with the Max Havelaar Foundation's labeling program in the Netherlands. In 2004, 24,222 metric tons (of 7,050,000 produced worldwide) were fair trade; in 2005, 33,991 metric tons out of 6,685,000 were fair trade, an increase from 0.34% to 0.51%. A number of fair trade impact studies have shown that fair trade coffee produces a mixed impact on the communities that grow it. Many studies are skeptical about fair trade, reporting that it often worsens the bargaining power of those who are not part of it. Coffee was incorporated into the fair-trade movement in 1988, when the Max Havelaar mark was introduced in the Netherlands. The very first fair-trade coffee was an effort to import a Guatemalan coffee into Europe as "Indio Solidarity Coffee".

Since the founding of organisations such as the European Fair Trade Association (1987), the production and consumption of fair trade coffee has grown as some local and national coffee chains started to offer fair trade alternatives. For example, in April 2000, after a year-long campaign by the human rights organization Global Exchange, Starbucks decided to carry fair-trade coffee in its stores. Since September 2009 all Starbucks Espresso beverages in UK and Ireland are made with Fairtrade and Shared Planet certified coffee.

A 2005 study done in Belgium concluded that consumers' buying behavior is not consistent with their positive attitude toward ethical products. On average 46% of European consumers claimed to be willing to pay substantially more for ethical products, including fair-trade products such as coffee. The study found that the majority of respondents were unwilling to pay the actual price premium of 27% for fair trade coffee.

Folklore and Culture

The Oromo people would customarily plant a coffee tree on the graves of powerful sorcerers. They believed that the first coffee bush sprang up from the tears that the god of heaven shed over the corpse of a dead sorcerer.

Johann Sebastian Bach was inspired to compose the *Coffee Cantata*, about dependence on the beverage.

Economic Impacts

Market volatility, and thus increased returns, during 1830 encouraged Brazilian entrepreneurs to shift their attention from gold to coffee, a crop hitherto reserved for local consumption. Concurrent with this shift was the commissioning of vital infrastructures, including approximately 7,000 km of railroads between 1860 and 1885. The creation of these railways enabled the importation of workers, in order to meet the enormous need for labor. This development primarily affected the State of Rio de Janeiro, as well as the Southern States of Brazil, most notably São Paulo, due to its favourable climate, soils, and terrain.

Coffee production attracted immigrants in search of better economic opportunities in the early 1900s. Mainly, these were Portuguese, Italian, Spanish, German, and Japanese nationals. For instance, São Paulo received approximately 733,000 immigrants in the decade preceding 1900, whilst only receiving approximately 201,000 immigrants in the six years to 1890. The production yield of coffee increases. In 1880, São Paulo produced 1.2 million bags (25% of total production), in

1888 2.6 million (40%), in 1902 8 million bags (60%). Coffee is then 63% of the country's exports. The gains made by this trade allow sustained economic growth in the country.

Map of coffee areas in Brazil

The four years between planting a coffee and the first harvest extends seasonal variations in the price of coffee. The Brazilian Government is thus forced, to some extent, to keep strong price subsidies during production periods.

Competition

Coffee competitions take place across the globe with people at the regional competing to achieve national titles and then compete on the international stage. World Coffee Events holds the largest of such events moving the location of the final competition each year. The competition includes the following events: Barista Championship, Brewers Cup, Latte Art and Cup Tasters. A World Brewer's Cup Championship takes place in Melbourne, Australia, every year that houses contestants from around the world to crown the World's Coffee King.

Fiber Crop

Fiber crops are field crops grown for their fibres, which are traditionally used to make paper, cloth, or rope. The fibers may be chemically modified, like in viscose (used to make rayon and cellophane). In recent years materials scientists have begun exploring further use of these fibers in composite materials.

Fiber crops are generally harvestable after a single growing season, as distinct from trees, which are typically grown for many years before being harvested for such materials as wood pulp fiber or lacebark. In specific circumstances, fiber crops can be superior to wood pulp fiber in terms of technical performance, environmental impact or cost.

There are a number of issues regarding the use of fiber crops to make pulp. One of these is seasonal availability. While trees can be harvested continuously, many field crops are harvested once during the year and must be stored such that the crop doesn't rot over a period of many months. Considering that many pulp mills require several thousand tonnes of fiber source per day, storage of the fiber source can be a major issue.

Botanically, the fibers harvested from many of these plants are bast fibers; the fibers come from the phloem tissue of the plant. The other fiber crop fibers are seed padding, leaf fiber, or other parts of the plant.

Fiber Sources

Ancient Sanskrit on Hemp based Paper. Hemp Fiber was commonly used in the production of paper from 200 BCE to the Late 1800's.

Before the industrialisation of the paper production the most common fibre source was recycled fibres from used textiles, called rags. The rags were from hemp, linen and cotton. A process for removing printing inks from recycled paper was invented by German jurist Justus Claproth in 1774. Today this method is called deinking. It was not until the introduction of wood pulp in 1843 that paper production was not dependent on recycled materials from ragpickers.

Fiber Crops

- Bast fibers (Stem-skin fibers)
  - Esparto, a fiber from a grass
  - Jute, widely used, it is the cheapest fiber after cotton
  - Flax, produces linen
  - Indian hemp, the Dogbane used by native Americans
  - Hemp, a soft, strong fiber, edible seeds
  - Hoopvine, also used for barrel hoops and baskets, edible leaves, medicine
  - Kenaf, the interior of the plant stem is used for its fiber. Edible leaves.
  - Linden Bast
  - Nettles
  - Ramie, a nettle, stronger than cotton or flax, makes "China grass cloth"
  - Papyrus, a pith fiber, akin to a bast fiber
- Leaf fibers
  - Abacá, a banana, producing "manila" rope from leaves
  - Sisal, often termed agave
  - Bowstring Hemp, an old use of a common decorative agave, also Sansevieria rox-

burghiana, Sansevieria hyacinthoides

- o Henequen, an agave. A useful fiber, but not as high quality as sisal
- o Phormium, "New Zealand Flax"
- o Yucca, an agave
- Seed fibers and fruit fibers
  - o Coir, the fiber from the coconut husk
  - o Cotton
  - o Kapok
  - o Milkweed, grown for the filament-like pappus in its seed pods
  - o Luffa, a gourd which when mature produces a sponge-like mass of xylem, used to make loofa sponge
- Other fibers (Leaf, fruit, and other fibers)
  - o Bamboo fiber, a viscose fiber like rayon, technically a semi-synthetic fiber

Fiber Dimensions

Source of pulp	Fiber length, mm	Fiber diameter, μm
Softwood	3.1	30
Hardwood	1.0	16
Wheat straw	1.5	13
Rice straw	1.5	9
Esparto grass	1.1	10
Reed	1.5	13
Bagasse	1.7	20
Bamboo	2.7	14
Cotton	25.0	20

Example of Fiber crop

Cotton

Manually decontaminating cotton before processing at an Indian spinning mill (2010)

Cotton is a soft, fluffy staple fiber that grows in a boll, or protective case, around the seeds of cotton plants of the genus *Gossypium* in the family of *Malvaceae*. The fiber is almost pure cellulose. Under natural conditions, the cotton bolls will tend to increase the dispersal of the seeds.

The plant is a shrub native to tropical and subtropical regions around the world, including the

Americas, Africa, and India. The greatest diversity of wild cotton species is found in Mexico, followed by Australia and Africa. Cotton was independently domesticated in the Old and New Worlds.

The fiber is most often spun into yarn or thread and used to make a soft, breathable textile. The use of cotton for fabric is known to date to prehistoric times; fragments of cotton fabric dated from 5000 BC have been excavated in Mexico and the Indus Valley Civilization in Indian subcontinent between 6000 BC and 5000 BC. Although cultivated since antiquity, it was the invention of the cotton gin that lowered the cost of production that led to its widespread use, and it is the most widely used natural fiber cloth in clothing today.

Current estimates for world production are about 25 million tonnes or 110 million bales annually, accounting for 2.5% of the world's arable land. China is the world's largest producer of cotton, but most of this is used domestically. The United States has been the largest exporter for many years. In the United States, cotton is usually measured in bales, which measure approximately 0.48 cubic meters (17 cubic feet) and weigh 226.8 kilograms (500 pounds).

Types

There are four commercially grown species of cotton, all domesticated in antiquity:

- *Gossypium hirsutum* – upland cotton, native to Central America, Mexico, the Caribbean and southern Florida (90% of world production)
- *Gossypium barbadense* – known as extra-long staple cotton, native to tropical South America (8% of world production)
- *Gossypium arboreum* – tree cotton, native to India and Pakistan (less than 2%)
- *Gossypium herbaceum* – Levant cotton, native to southern Africa and the Arabian Peninsula (less than 2%)

The two New World cotton species account for the vast majority of modern cotton production, but the two Old World species were widely used before the 1900s. While cotton fibers occur naturally in colors of white, brown, pink and green, fears of contaminating the genetics of white cotton have led many cotton-growing locations to ban the growing of colored cotton varieties, which remain a specialty product.

History

Indian Subcontinent

The earliest evidence of cotton use in the Indian subcontinent has been found at the site of Mehrgarh and Rakhigarhi where cotton threads have been found preserved in copper beads; these finds have been dated to Neolithic (between 6000 and 5000 BC). Cotton cultivation in the region is dated to the Indus Valley Civilization, which covered parts of modern eastern Pakistan and northwestern India between 3300 and 1300 BC. The Indus cotton industry was well-developed and some methods used in cotton spinning and fabrication continued to be used until the industrialization of India. Between 2000 and 1000 BC cotton became widespread across much of India. For example, it has been found at the site of Hallus in Karnataka dating from around 1000 BC.

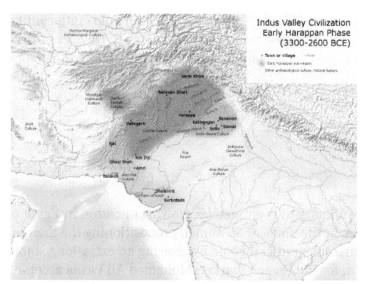

Indus Valley Civilization, Early Phase (3300-2600 BC)

Mexico

Cotton fabrics discovered in a cave near Tehuacán, Mexico have been dated to around 5800 BC. The ***domestication*** of Gossypium hirsutum in Mexico is dated between 3400 and 2300 BC.

Peru

In Peru, cultivation of the indigenous cotton species *Gossypium barbadense* has been dated, from a find in Ancon, to c 4200 BC, and was the backbone of the development of coastal cultures such as the Norte Chico, Moche, and Nazca. Cotton was grown upriver, made into nets, and traded with fishing villages along the coast for large supplies of fish. The Spanish who came to Mexico and Peru in the early 16th century found the people growing cotton and wearing clothing made of it.

Arabia

The Greeks and the Arabs were not familiar with cotton until the Wars of Alexander the Great, as his contemporary Megasthenes told Seleucus I Nicator of "there being trees on which wool grows" in "Indica". This may be a reference to "tree cotton", Gossypium arboreum, which is a native of the Indian subcontinent.

According to the *Columbia Encyclopedia*:

Cotton has been spun, woven, and dyed since prehistoric times. It clothed the people of ancient India, Egypt, and China. Hundreds of years before the Christian era, cotton textiles were woven in India with matchless skill, and their use spread to the Mediterranean countries.

Iran

In Iran (Persia), the history of cotton dates back to the Achaemenid era (5th century BC); however, there are few sources about the planting of cotton in pre-Islamic Iran. The planting of cotton was common in Merv, Ray and Pars of Iran. In Persian poets' poems, especially Ferdowsi's Shahname, there are references to cotton ("panbe" in Persian). Marco Polo (13th century) refers to the major products of Persia, including cotton. John Chardin, a French traveler of the 17th century who visited the Safavid Persia, spoke approvingly of the vast cotton farms of Persia.

China

During the Han dynasty (207 BC - 220 AD), cotton was grown by Chinese peoples in the southern Chinese province of Yunnan.

Egypt

Though known since antiquity the commercial growing of cotton in Egypt only started in 1820's, following a Frenchman, by the name of M. Jumel, propositioning the then ruler, Mohamed Ali Pasha, that he could earn a substantial income by growing an extra-long staple Maho (Barbadence) cotton, in Lower Egypt, for the French market. Mohamed Ali Pasha accepted the proposition and granted himself the monopoly on the sale and export of cotton in Egypt; and later dictated cotton should be grown in preference to other crops. By the time of the American Civil war annual exports had reached $16 million (120,000 bales), which rose to $56 million by 1864, primarily due to the loss of the Confederate supply on the world market. Exports continued to grow even after the reintroduction of US cotton, produced now by a paid workforce, and Egyptian exports reached 1.2 million bales a year by 1903.

Europe

Cotton plants as imagined and drawn by John Mandeville in the 14th century

During the late medieval period, cotton became known as an imported fiber in northern Europe, without any knowledge of how it was derived, other than that it was a plant. Because Herodotus had written in his *Histories*, Book III, 106, that in India trees grew in the wild producing wool, it was assumed that the plant was a tree, rather than a shrub. This aspect is retained in the name for cotton in several Germanic languages, such as German *Baumwolle*, which translates as "tree wool" (*Baum* means "tree"; *Wolle* means "wool"). Noting its similarities to wool, people in the region could only imagine that cotton must be produced by plant-borne sheep. John Mandeville, writing in 1350, stated as fact the now-preposterous belief: "There grew there [India] a wonderful tree which bore tiny lambs on the endes of its branches. These branches were so pliable that they bent down to allow the lambs to feed when they are hungrie [*sic*]." (See Vegetable Lamb of Tartary.) By the end of the 16th century, cotton was cultivated throughout the warmer regions in Asia and the Americas.

The Vegetable Lamb of Tartary

India's cotton-processing sector gradually declined during British expansion in India and the establishment of colonial rule during the late 18th and early 19th centuries. This was largely due to aggressive colonialist mercantile policies of the British East India Company, which made cotton processing and manufacturing workshops in India uncompetitive. Indian markets were increasingly forced to supply only raw cotton and, by British-imposed law, to purchase manufactured textiles from Britain.

Industrial Revolution in Britain

The advent of the Industrial Revolution in Britain provided a great boost to cotton manufacture, as textiles emerged as Britain's leading export. In 1738, Lewis Paul and John Wyatt, of Birmingham, England, patented the roller spinning machine, as well as the flyer-and-bobbin system for drawing cotton to a more even thickness using two sets of rollers that traveled at different speeds. Later, the invention of the James Hargreaves' spinning jenny in 1764, Richard Arkwright's spinning frame in 1769 and Samuel Crompton's spinning mule in 1775 enabled British spinners to produce cotton yarn at much higher rates. From the late 18th century on, the British city of Manchester acquired the nickname *"Cottonopolis"* due to the cotton industry's omnipresence within the city, and Manchester's role as the heart of the global cotton trade.

Production capacity in Britain and the United States was improved by the invention of the cotton gin by the American Eli Whitney in 1793. Before the development of cotton gins, the cotton fibers had to be pulled from the seeds tediously by hand. By the late 1700s a number of crude ginning machines had been developed. However, to produce a bale of cotton required over 600 hours of human labor, making large-scale production uneconomical in the United States, even with the use of humans as slave labor. The gin that Whitney manufactured (the Holmes design) reduced the hours down to just a dozen or so per bale. Although Whitney patented his own design for a cotton gin, he manufactured a prior design from Henry Odgen Holmes, for which Holmes filed a patent in 1796. Improving technology and increasing control of world markets allowed British traders to develop a commercial chain in which raw cotton fibers were (at first) purchased from colonial plantations, processed into cotton cloth in the mills of Lancashire, and then exported on British ships to captive colonial markets in West Africa, India, and China (via Shanghai and Hong Kong).

By the 1840s, India was no longer capable of supplying the vast quantities of cotton fibers needed by mechanized British factories, while shipping bulky, low-price cotton from India to Britain was time-consuming and expensive. This, coupled with the emergence of American cotton as a superior type (due to the longer, stronger fibers of the two domesticated native American species, *Gossypium hirsutum* and *Gossypium barbadense*), encouraged British traders to purchase cotton from plantations in the United States and plantations in the Caribbean. By the mid-19th century, "King Cotton" had become the backbone of the southern American economy. In the United States, cultivating and harvesting cotton became the leading occupation of slaves.

During the American Civil War, American cotton exports slumped due to a Union blockade on Southern ports, and also because of a strategic decision by the Confederate government to cut exports, hoping to force Britain to recognize the Confederacy or enter the war. This prompted the main purchasers of cotton, Britain and France, to turn to Egyptian cotton. British and French traders invested heavily in cotton plantations. The Egyptian government of Viceroy Isma'il took out substantial loans from European bankers and stock exchanges. After the American Civil War ended in 1865, British and French traders abandoned Egyptian cotton and returned to cheap American exports, sending Egypt into a deficit spiral that led to the country declaring bankruptcy in 1876, a key factor behind Egypt's occupation by the British Empire in 1882.

During this time, cotton cultivation in the British Empire, especially Australia and India, greatly increased to replace the lost production of the American South. Through tariffs and other restrictions, the British government discouraged the production of cotton cloth in India; rather, the raw fiber was sent to England for processing. The Indian Mahatma Gandhi described the process:

1. English people buy Indian cotton in the field, picked by Indian labor at seven cents a day, through an optional monopoly.

2. This cotton is shipped on British ships, a three-week journey across the Indian Ocean, down the Red Sea, across the Mediterranean, through Gibraltar, across the Bay of Biscay and the Atlantic Ocean to London. One hundred per cent profit on this freight is regarded as small.

3. The cotton is turned into cloth in Lancashire. You pay shilling wages instead of Indian pennies to your workers. The English worker not only has the advantage of better wages, but the steel companies of England get the profit of building the factories and machines. Wages; profits; all these are spent in England.

4. The finished product is sent back to India at European shipping rates, once again on British ships. The captains, officers, sailors of these ships, whose wages must be paid, are English.

The only Indians who profit are a few lascars who do the dirty work on the boats for a few cents a day.

5. The cloth is finally sold back to the kings and landlords of India who got the money to buy this expensive cloth out of the poor peasants of India who worked at seven cents a day.

United States

Prisoners farming cotton under the trusty system in Parchman Farm, Mississippi, 1911

In the United States, Southern cotton provided capital for the continuing development of the North. The cotton produced by enslaved African Americans not only helped the South, but also enriched Northern merchants. Much of the Southern cotton was trans-shipped through northern ports.

Cotton remained a key crop in the Southern economy after emancipation and the end of the Civil War in 1865. Across the South, sharecropping evolved, in which landless black and white farmers worked land owned by others in return for a share of the profits. Some farmers rented the land and bore the production costs themselves. Until mechanical cotton pickers were developed, cotton farmers needed additional labor to hand-pick cotton. Picking cotton was a source of income for families across the South. Rural and small town school systems had split vacations so children could work in the fields during "cotton-picking."

It was not until the 1950s that reliable harvesting machinery was introduced (prior to this, cotton-harvesting machinery had been too clumsy to pick cotton without shredding the fibers). During the first half of the 20th century, employment in the cotton industry fell, as machines began to replace laborers and the South's rural labor force dwindled during the World Wars.

Cotton remains a major export of the southern United States, and a majority of the world's annual cotton crop is of the long-staple American variety.

Cultivation

Successful cultivation of cotton requires a long frost-free period, plenty of sunshine, and a moderate rainfall, usually from 60 to 120 cm (24 to 47 in). Soils usually need to be fairly heavy, although the level of nutrients does not need to be exceptional. In general, these conditions are met within the seasonally dry tropics and subtropics in the Northern and Southern hemispheres, but a large proportion of the cotton grown today is cultivated in areas with less rainfall that obtain the water from irrigation. Production of the crop for a given year usually starts soon after harvesting the preceding autumn. Cotton is naturally a perennial but is grown as an annual to help control pests. Planting time in spring in the Northern hemisphere varies from the beginning of February to the beginning of June. The area of the United States known as the South Plains is the largest contiguous cotton-growing region in the world. While dryland (non-irrigated) cotton is successfully grown in this region, consistent yields are only produced with heavy reliance on irrigation water drawn from the Ogallala Aquifer. Since cotton is somewhat salt and drought tolerant, this makes it an attractive crop for arid and semiarid regions. As water resources get tighter around the world, economies that rely on it face difficulties and conflict, as well as potential environmental problems. For example, improper cropping and irrigation practices have led to desertification in areas of Uzbekistan, where cotton is a major export. In the days of the Soviet Union, the Aral Sea was tapped for agricultural irrigation, largely of cotton, and now salination is widespread.

Cotton field

Cotton plant

A cotton field, late in the season

Cotton plowing in Togo, 1928

Picking cotton in Armenia in the 1930s. No cotton is grown there today.

Cotton ready for shipment, Houston, Texas (postcard, circa 1911)

Cotton modules in Australia (2007)

Cotton can also be cultivated to have colors other than the yellowish off-white typical of modern commercial cotton fibers. Naturally colored cotton can come in red, green, and several shades of brown.

Genetic Modification

Genetically modified (GM) cotton was developed to reduce the heavy reliance on pesticides. The bacterium *Bacillus thuringiensis* (Bt) naturally produces a chemical harmful only to a small fraction of insects, most notably the larvae of moths and butterflies, beetles, and flies, and harmless to other forms of life. The gene coding for Bt toxin has been inserted into cotton, causing cotton, called Bt cotton, to produce this natural insecticide in its tissues. In many regions, the main pests in commercial cotton are lepidopteran larvae, which are killed by the Bt protein in the transgenic cotton they eat. This eliminates the need to use large amounts of broad-spectrum insecticides to kill lepidopteran pests (some of which have developed pyrethroid resistance). This spares natural insect predators in the farm ecology and further contributes to noninsecticide pest management.

But Bt cotton is ineffective against many cotton pests, however, such as plant bugs, stink bugs, and aphids; depending on circumstances it may still be desirable to use insecticides against these. A 2006 study done by Cornell researchers, the Center for Chinese Agricultural Policy and the Chinese Academy of Science on Bt cotton farming in China found that after seven years these secondary pests that were normally controlled by pesticide had increased, necessitating the use of pesticides at similar levels to non-Bt cotton and causing less profit for farmers because of the extra expense of GM seeds. However, a 2009 study by the Chinese Academy of Sciences, Stanford University and Rutgers University refuted this. They concluded that the GM cotton effectively controlled bollworm. The secondary pests were mostly miridae (plant bugs) whose increase was related to local temperature and rainfall and only continued to increase in half the villages studied. Moreover, the increase in insecticide use for the control of these secondary insects was far smaller than the reduction in total insecticide use due to Bt cotton adoption. A 2012 Chinese study concluded that Bt cotton halved the use of pesticides and doubled the level of ladybirds, lacewings and spiders. The International Service for the Acquisition of Agri-biotech Applications (ISAAA) said that, worldwide, GM cotton was planted on an area of 25 million hectares in 2011. This was 69% of the worldwide total area planted in cotton.

GM cotton acreage in India grew at a rapid rate, increasing from 50,000 hectares in 2002 to 10.6 million hectares in 2011. The total cotton area in India was 12.1 million hectares in 2011, so GM cotton was grown on 88% of the cotton area. This made India the country with the largest area of GM cotton in the world. A long-term study on the economic impacts of Bt cotton in India, published in the Journal PNAS in 2012, showed that Bt cotton has increased yields, profits, and living standards of smallholder farmers. The U.S. GM cotton crop was 4.0 million hectares in 2011 the second largest area in the world, the Chinese GM cotton crop was third largest by area with 3.9 million hectares and Pakistan had the fourth largest GM cotton crop area of 2.6 million hectares in 2011. The initial introduction of GM cotton proved to be a success in Australia – the yields were equivalent to the non-transgenic varieties and the crop used much less pesticide to produce (85% reduction). The subsequent introduction of a second variety of GM cotton led to increases in GM cotton production until 95% of the Australian cotton crop was GM in 2009 making Australia the country with the fifth largest GM cotton crop in the world. Other GM cotton growing countries in 2011 were Argentina, Myanmar, Burkina Faso, Brazil, Mexico, Colombia, South Africa and Costa Rica.

Cotton has been genetically modified for resistance to glyphosate a broad-spectrum herbicide discovered by Monsanto which also sells some of the Bt cotton seeds to farmers. There are also a number of other cotton seed companies selling GM cotton around the world. About 62% of the GM cotton grown from 1996 to 2011 was insect resistant, 24% stacked product and 14% herbicide resistant.

Cotton has gossypol, a toxin that makes it inedible. However, scientists have silenced the gene that produces the toxin, making it a potential food crop.

Organic Production

Organic cotton is generally understood as cotton from plants not genetically modified and that is certified to be grown without the use of any synthetic agricultural chemicals, such as fertilizers or pesticides. Its production also promotes and enhances biodiversity and biological cycles. In the United States, organic cotton plantations are required to enforce the National Organic Program (NOP). This institution determines the allowed practices for pest control, growing, fertilizing, and handling of organic crops. As of 2007, 265,517 bales of organic cotton were produced in 24 countries, and worldwide production was growing at a rate of more than 50% per year.

Pests and Weeds

Hoeing a cotton field to remove weeds, Greene County, Georgia, US, 1941

The cotton industry relies heavily on chemicals, such as herbicides, fertilizers and insecticides, although a very small number of farmers are moving toward an organic model of production, and organic cotton products are now available for purchase at limited locations. These are popular for baby clothes and diapers. Under most definitions, organic products do not use genetic engineering. All natural cotton products are known to be both sustainable and hypoallergenic.

Historically, in North America, one of the most economically destructive pests in cotton production has been the boll weevil. Due to the US Department of Agriculture's highly successful Boll Weevil Eradication Program (BWEP), this pest has been eliminated from cotton in most of the United States. This program, along with the introduction of genetically engineered Bt cotton (which contains a bacterial gene that codes for a plant-produced protein that is toxic to a number of pests such as cotton bollworm and pink bollworm), has allowed a reduction in the use of synthetic insecticides.

Female and nymph Cotton Harlequin Bug

Other significant global pests of cotton include the pink bollworm, *Pectinophora gossypiella*; the chili thrips, *Scirtothrips dorsalis*; the cotton seed bug, *Oxycarenus hyalinipennis*; the tarnish plant bug, *Lygus lineolaris*; and the fall armyworm, *Spodoptera frugiperda, Xanthomonas citri subsp. malvacearum*.

Harvesting

Most cotton in the United States, Europe and Australia is harvested mechanically, either by a cotton picker, a machine that removes the cotton from the boll without damaging the cotton plant, or by a cotton stripper, which strips the entire boll off the plant. Cotton strippers are used in regions where it is too windy to grow picker varieties of cotton, and usually after application of a chemical defoliant or the natural defoliation that occurs after a freeze. Cotton is a perennial crop in the tropics, and without defoliation or freezing, the plant will continue to grow.

Cotton continues to be picked by hand in developing countries.

Offloading freshly harvested cotton into a module builder in Texas; previously
built modules can be seen in the background

Cotton being picked by hand in India, 2005.

Competition from Synthetic Fibers

The era of manufactured fibers began with the development of rayon in France in the 1890s. Rayon is derived from a natural cellulose and cannot be considered synthetic, but requires extensive processing in a manufacturing process, and led the less expensive replacement of more naturally derived materials. A succession of new synthetic fibers were introduced by the chemicals industry in the following decades. Acetate in fiber form was developed in 1924. Nylon, the first fiber synthesized entirely from petrochemicals, was introduced as a sewing thread by DuPont in 1936, followed by DuPont's acrylic in 1944. Some garments were created from fabrics based on these fibers, such as women's hosiery from nylon, but it was not until the introduction of polyester into the fiber marketplace in the early 1950s that the market for cotton came under threat. The rapid uptake of polyester garments in the 1960s caused economic hardship in cotton-exporting economies, especially in Central American countries, such as Nicaragua, where cotton production had boomed tenfold between 1950 and 1965 with the advent of cheap chemical pesticides. Cotton production recovered in the 1970s, but crashed to pre-1960 levels in the early 1990s.

Uses

Cotton is used to make a number of textile products. These include terrycloth for highly absorbent bath towels and robes; denim for blue jeans; cambric, popularly used in the manufacture of blue work shirts (from which we get the term "blue-collar"); and corduroy, seersucker, and cotton twill. Socks, underwear, and most T-shirts are made from cotton. Bed sheets often are made from cotton. Cotton also is used to make yarn used in crochet and knitting. Fabric also can be made from recycled or recovered cotton that otherwise would be thrown away during the spinning, weaving, or cutting process. While many fabrics are made completely of cotton, some materials blend cotton with other fibers, including rayon and synthetic fibers such as polyester. It can either be used in knitted or woven fabrics, as it can be blended with elastine to make a stretchier thread for knitted fabrics, and apparel such as stretch jeans.

In addition to the textile industry, cotton is used in fishing nets, coffee filters, tents, explosives manufacture, cotton paper, and in bookbinding. The first Chinese paper was made of cotton fiber. Fire hoses were once made of cotton.

The cottonseed which remains after the cotton is ginned is used to produce cottonseed oil, which, after refining, can be consumed by humans like any other vegetable oil. The cottonseed meal that is left generally is fed to ruminant livestock; the gossypol remaining in the meal is toxic to monogastric animals. Cottonseed hulls can be added to dairy cattle rations for roughage. During the American slavery period, cotton root bark was used in folk remedies as an abortifacient, that is, to induce a miscarriage. Gossypol was one of the many substances found in all parts of the cotton plant and it was described by the scientists as 'poisonous pigment'. It also appears to inhibit the development of sperm or even restrict the mobility of the sperm. Also, it is thought to interfere with the menstrual cycle by restricting the release of certain hormones.

Cotton linters are fine, silky fibers which adhere to the seeds of the cotton plant after ginning. These curly fibers typically are less than $\frac{1}{8}$ inch (3.2 mm) long. The term also may apply to the longer textile fiber staple lint as well as the shorter fuzzy fibers from some upland species. Linters are traditionally used in the manufacture of paper and as a raw material in the manufacture of cellulose. In the UK, linters are referred to as "cotton wool". This can also be a refined product (*absorbent cotton* in U.S. usage) which has medical, cosmetic and many other practical uses. The first medical use of cotton wool was by Sampson Gamgee at the Queen's Hospital (later the General Hospital) in Birmingham, England.

Shiny cotton is a processed version of the fiber that can be made into cloth resembling satin for shirts and suits. However, it is hydrophobic (does not absorb water easily), which makes it unfit for use in bath and dish towels (although examples of these made from shiny cotton are seen).

Conton in a tree

The name Egyptian cotton is broadly associated with quality products, however only a small percentage of "Egyptian cotton" products are actually of superior quality. Most products bearing the name are not made with cotton from Egypt.

Pima cotton is often compared to Egyptian cotton, as both are used in high quality bed sheets and other cotton products. It is considered the next best quality after high quality Egyptian cotton by some authorities. Pima cotton is grown in the American southwest. Not all products bearing the

Pima name are made with the finest cotton. The Pima name is now used by cotton-producing nations such as Peru, Australia and Israel.

Cotton lisle is a finely-spun, tightly twisted type of cotton that is noted for being strong and durable. Lisle is composed of two strands that have each been twisted an extra twist per inch than ordinary yarns and combined to create a single thread. The yarn is spun so that it is compact and solid. This cotton is used mainly for underwear, stockings, and gloves. Colors applied to this yarn are noted for being more brilliant than colors applied to softer yarn. This type of thread was first made in the city of Lisle, France (now Lille), hence its name.

International Trade

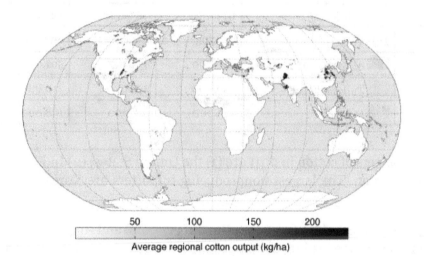

Average regional cotton output (kg/ha)

Worldwide cotton production

The largest producers of cotton, currently (2009), are China and India, with annual production of about 34 million bales and 33.4 million bales, respectively; most of this production is consumed by their respective textile industries. The largest exporters of raw cotton are the United States, with sales of $4.9 billion, and Africa, with sales of $2.1 billion. The total international trade is estimated to be $12 billion. Africa's share of the cotton trade has doubled since 1980. Neither area has a significant domestic textile industry, textile manufacturing having moved to developing nations in Eastern and South Asia such as India and China. In Africa, cotton is grown by numerous small holders. Dunavant Enterprises, based in Memphis, Tennessee, is the leading cotton broker in Africa, with hundreds of purchasing agents. It operates cotton gins in Uganda, Mozambique, and Zambia. In Zambia, it often offers loans for seed and expenses to the 180,000 small farmers who grow cotton for it, as well as advice on farming methods. Cargill also purchases cotton in Africa for export.

The 25,000 cotton growers in the United States are heavily subsidized at the rate of $2 billion per year although China now provides the highest overall level of cotton sector support. The future of these subsidies is uncertain and has led to anticipatory expansion of cotton brokers' operations in Africa. Dunavant expanded in Africa by buying out local operations. This is only possible in former British colonies and Mozambique; former French colonies continue to maintain tight monopolies, inherited from their former colonialist masters, on cotton purchases at low fixed prices.

Leading Producer Countries

Top 10 Cotton Producing Countries (in metric tonnes)				
Rank	Country	2010	2012	2014
1	China	5,970,000	6,281,000	6,532,000
2	India	5,683,000	6,071,000	6,423,000
3	United States	3,941,700	3,412,550	3,553,000
4	Pakistan	1,869,000	2,312,000	2,308,000
5	Brazil	973,449	1,673,337	1,524,103
6	Uzbekistan	1,136,120	983,400	849,000
7	Turkey	816,705	754,600	697,000
8	Australia	386,800	473,497	501,000
9	Turkmenistan	230,000	295,000	210,000
10	Mexico	225,000	195,000	198,000

The five leading exporters of cotton in 2011 are (1) the United States, (2) India, (3) Brazil, (4) Australia, and (5) Uzbekistan. The largest nonproducing importers are Korea, Taiwan, Russia, and Japan.

In India, the states of Maharashtra (26.63%), Gujarat (17.96%) and Andhra Pradesh (13.75%) and also Madhya Pradesh are the leading cotton producing states, these states have a predominantly tropical wet and dry climate.

In the United States, the state of Texas led in total production as of 2004, while the state of California had the highest yield per acre.

Fair Trade

Cotton is an enormously important commodity throughout the world. However, many farmers in developing countries receive a low price for their produce, or find it difficult to compete with developed countries.

This has led to an international dispute:

On 27 September 2002, Brazil requested consultations with the US regarding prohibited and actionable subsidies provided to US producers, users and/or exporters of upland cotton, as well as legislation, regulations, statutory instruments and amendments thereto providing such subsidies (including export credits), grants, and any other assistance to the US producers, users and exporters of upland cotton.

On 8 September 2004, the Panel Report recommended that the United States "withdraw" export credit guarantees and payments to domestic users and exporters, and "take appropriate steps to remove the adverse effects or withdraw" the mandatory price-contingent subsidy measures.

While Brazil was fighting the US through the WTO's Dispute Settlement Mechanism against a heavily subsidized cotton industry, a group of four least-developed African countries – Benin, Burkina Faso, Chad, and Mali – also known as "Cotton-4" have been the leading protagonist for the reduction of US cotton subsidies through negotiations. The four introduced a "Sectoral Initiative in Favour of Cotton", presented by Burkina Faso's President Blaise Compaoré during the Trade Negotiations Committee on 10 June 2003.

In addition to concerns over subsidies, the cotton industries of some countries are criticized for employing child labor and damaging workers' health by exposure to pesticides used in production. The Environmental Justice Foundation has campaigned against the prevalent use of forced child and adult labor in cotton production in Uzbekistan, the world's third largest cotton exporter. The international production and trade situation has led to "fair trade" cotton clothing and footwear, joining a rapidly growing market for organic clothing, fair fashion or "ethical fashion". The fair trade system was initiated in 2005 with producers from Cameroon, Mali and Senegal.

Trade

Cotton is bought and sold by investors and price speculators as a tradable commodity on 2 different stock exchanges in the United States of America.

- Cotton No. 2 futures contracts are traded on the New York Board of Trade (NYBOT) under the ticker symbol CT. They are delivered every year in March, May, July, October, and December.

- Cotton futures contracts are traded on the New York Mercantile Exchange (NYMEX) under the ticker symbol TT. They are delivered every year in March, May, July, October, and December.

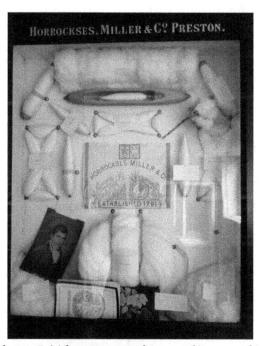

A display from a British cotton manufacturer of items used in a cotton
mill during the Industrial Revolution.

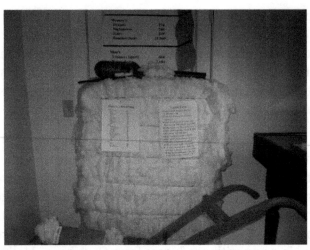

A bale of cotton on display at the Louisiana State Cotton Museum in Lake Providence in East Carroll Parish in northeastern Louisiana

Critical Temperatures

- Favorable travel temperature range: below 25 °C (77 °F)

- Optimum travel temperature: 21 °C (70 °F)

- Glow temperature: 205 °C (401 °F)

- Fire point: 210 °C (410 °F)

- Autoignition temperature: 360 °C (680 °F) - 425 °C (797 °F)

- Autoignition temperature (for oily cotton): 120 °C (248 °F)

A temperature range of 25 to 35 °C (77 to 95 °F) is the optimal range for mold development. At temperatures below 0 °C (32 °F), rotting of wet cotton stops. Damaged cotton is sometimes stored at these temperatures to prevent further deterioration.

British Standard Yarn Measures

- 1 thread = 55 in or 140 cm

- 1 skein or rap = 80 threads (120 yd or 110 m)

- 1 hank = 7 skeins (840 yd or 770 m)

- 1 spindle = 18 hanks (15,120 yd or 13.83 km)

Fiber Properties

Property	Evaluation
Shape	Fairly uniform in width, 12–20 micrometers; length varies from 1 cm to 6 cm (½ to 2½ inches); typical length is 2.2 cm to 3.3 cm (⅞ to 1¼ inches).

Luster	high
Tenacity (strength) Dry Wet	 3.0–5.0 g/d 3.3–6.0 g/d
Resiliency	low
Density	1.54–1.56 g/cm³
Moisture absorption raw: conditioned saturation mercerized: conditioned saturation	 8.5% 15–25% 8.5–10.3% 15–27%+
Dimensional stability	good
Resistance to acids alkali organic solvents sunlight microorganisms insects	 damage, weaken fibers resistant; no harmful effects high resistance to most Prolonged exposure weakens fibers. Mildew and rot-producing bacteria damage fibers. Silverfish damage fibers.
Thermal reactions to heat to flame	 Decomposes after prolonged exposure to temperatures of 150 °C or over. Burns readily.

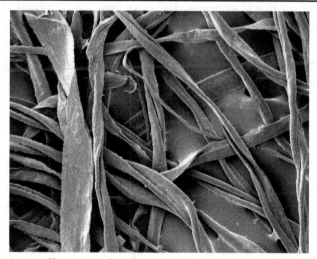

Cotton fibers viewed under a scanning electron microscope

The chemical composition of cotton is as follows:

- cellulose 91.00%

- water 7.85%

- protoplasm, pectins 0.55%

- waxes, fatty substances 0.40%

- mineral salts 0.20%

Cotton Genome

A public genome sequencing effort of cotton was initiated in 2007 by a consortium of public researchers. They agreed on a strategy to sequence the genome of cultivated, tetraploid cotton. "Tetraploid" means that cultivated cotton actually has two separate genomes within its nucleus, referred to as the A and D genomes. The sequencing consortium first agreed to sequence the D-genome relative of cultivated cotton (*G. raimondii*, a wild Central American cotton species) because of its small size and limited number of repetitive elements. It is nearly one-third the number of bases of tetraploid cotton (AD), and each chromosome is only present once. The A genome of *G. arboreum* would be sequenced next. Its genome is roughly twice the size of *G. raimondii*'s. Part of the difference in size between the two genomes is the amplification of *retrotransposons* (GORGE). Once both diploid genomes are assembled, then research could begin sequencing the actual genomes of cultivated cotton varieties. This strategy is out of necessity; if one were to sequence the tetraploid genome without model diploid genomes, the euchromatic DNA sequences of the AD genomes would co-assemble and the repetitive elements of AD genomes would assembly independently into A and D sequences respectively. Then there would be no way to untangle the mess of AD sequences without comparing them to their diploid counterparts.

The public sector effort continues with the goal to create a high-quality, draft genome sequence from reads generated by all sources. The public-sector effort has generated Sanger reads of BACs, fosmids, and plasmids as well as 454 reads. These later types of reads will be instrumental in assembling an initial draft of the D genome. In 2010, two companies (Monsanto and Illumina), completed enough Illumina sequencing to cover the D genome of *G. raimondii* about 50x. They announced that they would donate their raw reads to the public. This public relations effort gave them some recognition for sequencing the cotton genome. Once the D genome is assembled from all of this raw material, it will undoubtedly assist in the assembly of the AD genomes of cultivated varieties of cotton, but a lot of hard work remains.

Coir

Coir or coconut fibre, is a natural fibre extracted from the husk of coconut and used in products such as floor mats, doormats, brushes, and mattresses. Coir is the fibrous material found between the hard, internal shell and the outer coat of a coconut. Other uses of brown coir (made from ripe coconut) are in upholstery padding, sacking and horticulture. White coir, harvested from unripe coconuts, is used for making finer brushes, string, rope and fishing nets.

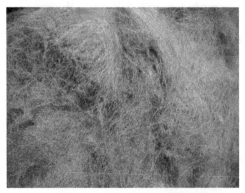

A close-up view of coir fibre

Segregation of coir fibre

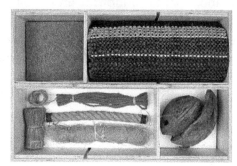

Various forms in which coir fibre can appear

Etymology

The English word "coir" comes from the Malayalam and Tamil word 'kayar' (കയി in Malayalam, கயிறு in Tamil).

History

Ropes and cordage made from coconut fibre have been in use from ancient times. Indian navigators who sailed the seas to Malaya, Java, China, and the Gulf of Arabia centuries ago used coir for their ship ropes. Arab writers of the 11th century AD referred to the extensive use of coir for ship ropes and rigging.

A coir industry in the UK was recorded before the second half of the 19th century. During 1840, Captain Widely, in co-operation with Captain Logan and Mr. Thomas Treloar, founded the known carpet firms of Treloar and Sons in Ludgate Hill, England, for the manufacture of coir into various fabrics suitable for floor coverings.

Structure

Coir fibres are found between the hard, internal shell and the outer coat of a coconut. The individual fibre cells are narrow and hollow, with thick walls made of cellulose. They are pale when immature, but later become hardened and yellowed as a layer of lignin is deposited on their walls. Each cell is about 1 mm (0.04 in) long and 10 to 20 μm (0.0004 to 0.0008 in) in diameter. Fibres are typically 10 to 30 centimetres (4 to 12 in) long. The two varieties of coir are brown and white. Brown coir harvested from fully ripened coconuts is thick, strong and has high abrasion resistance.

It is typically used in mats, brushes and sacking. Mature brown coir fibres contain more lignin and less cellulose than fibres such as flax and cotton, so are stronger but less flexible. White coir fibres harvested from coconuts before they are ripe are white or light brown in color and are smoother and finer, but also weaker. They are generally spun to make yarn used in mats or rope.

The coir fibre is relatively waterproof, and is one of the few natural fibres resistant to damage by saltwater. Fresh water is used to process brown coir, while seawater and fresh water are both used in the production of white coir.

Processing

Green coconuts, harvested after about six to 12 months on the palm, contain pliable white fibres. Brown fibre is obtained by harvesting fully mature coconuts when the nutritious layer surrounding the seed is ready to be processed into copra and desiccated coconut. The fibrous layer of the fruit is then separated from the hard shell (manually) by driving the fruit down onto a spike to split it (dehusking). A well-seasoned husker can manually separate 2,000 coconuts per day. Machines are now available which crush the whole fruit to give the loose fibres. These machines can process up to 2,000 coconuts per hour.

Brown Fibre

The fibrous husks are soaked in pits or in nets in a slow-moving body of water to swell and soften the fibres. The long bristle fibres are separated from the shorter mattress fibres underneath the skin of the nut, a process known as wet-milling. The mattress fibres are sifted to remove dirt and other rubbish, dried in the sun and packed into bales. Some mattress fibre is allowed to retain more moisture so it retains its elasticity for twisted fibre production. The coir fibre is elastic enough to twist without breaking and it holds a curl as though permanently waved. Twisting is done by simply making a rope of the hank of fibre and twisting it using a machine or by hand. The longer bristle fibre is washed in clean water and then dried before being tied into bundles or hanks. It may then be cleaned and 'hackled' by steel combs to straighten the fibres and remove any shorter fibre pieces. Coir bristle fibre can also be bleached and dyed to obtain hanks of different colours.

White Fibre

The immature husks are suspended in a river or water-filled pit for up to ten months. During this time, micro-organisms break down the plant tissues surrounding the fibres to loosen them — a process known as retting. Segments of the husk are then beaten by hand to separate out the long fibres which are subsequently dried and cleaned. Cleaned fibre is ready for spinning into yarn using a simple one-handed system or a spinning wheel.

Researchers at CSIR's National Institute for Interdisciplinary Science and Technology in Thiruvananthapuram have developed a biological process for the extraction of coir fibre from coconut husk without polluting the environment. The technology uses enzymes to separate the fibres by converting and solubilizing plant compounds to curb the pollution of waters caused by retting of husks.

Buffering

Because coir is high in sodium and potassium, it is treated before use as a growth medium for plants or fungi by soaking in a calcium buffering solution; most coir sold for growing purposes is pre-treated. Once any remaining salts have been leached out of the coir pith, it and the coir bark become suitable substrates for cultivating fungi. Coir is naturally rich in potassium, which can lead to magnesium deficiencies in soilless horticultural media.

Coir does provide a suitable substrate for horticultural use as a soilless potting medium. The material's high lignin content is longer lasting, holds more water, and does not shrink off the sides of the pot when dry allowing for easier rewetting. This light media has advantages and disadvantages that can be corrected with the addition of the proper amendment such as coarse sand for weight in interior plants like Draceana. Nutritive amendments should also be considered. Calcium and magnesium will be lacking in coir potting mixes, so a naturally good source of these nutrients is dolomitic lime which contains both. The addition of beneficial microbes to the coir media have been successful in tropical green house conditions and interior spaces as well. However, it is important to note that the microbes will engage in growth and reproduction under moist atmospheres producing fruiting bodies (mushrooms).

Bristle Coir

Bristle coir is the longest variety of coir fibre. It is manufactured from retted coconut husks through a process called defibring. The coir fibre thus extracted is then combed using steel combs to make the fibre clean and to remove short fibres. Bristle coir fibre is used as bristles in brushes for domestic and industrial applications.

Uses

Red coir is used in floor mats and doormats, brushes, mattresses, floor tiles and sacking. A small amount is also made into twine. Pads of curled brown coir fibre, made by needle-felting (a machine technique that mats the fibres together), are shaped and cut to fill mattresses and for use in erosion control on river banks and hillsides. A major proportion of brown coir pads are sprayed with rubber latex which bonds the fibres together (rubberised coir) to be used as upholstery padding for the automobile industry in Europe. The material is also used for insulation and packaging.

The major use of white coir is in rope manufacture. Mats of woven coir fibre are made from the finer grades of bristle and white fibre using hand or mechanical looms. White coir also is used to make fishing nets due to its strong resistance to saltwater.

In horticulture, coir is a substitute for sphagnum moss because it is free of bacteria and fungal spores. Coir is also useful to deter snails from delicate plantings, and as a growing medium in intensive glasshouse (greenhouse) horticulture.

Coconut coir from Mexico has been found to contain large numbers of colonies of the beneficial fungus *Aspergillus terreus*, which acts as a biological control against plant pathogenic fungi.

Making coir rope in Kerala, India

Coir is also used as a substrate to grow mushrooms. The coir is usually mixed with vermiculite and pasteurized with boiling water. After the coir/vermiculite mix has cooled to room temperature, it is placed in a larger container, usually a plastic box. Previously prepared spawn jars are then added, spawn is usually grown in jars using substrates such as rye grains or wild bird seed. This spawn is the mushrooms mycelium and will colonize the coir/vermiculite mix eventually fruiting mushrooms.

Coir is an allergen, as well as the latex and other materials used frequently in the treatment of coir.

Coir can be used as a terrarium substrate for reptiles or arachnids.

Major Producers

Total world coir fibre production is 250,000 tonnes (250,000 long tons; 280,000 short tons). This industry is particularly important in some areas of the developing world. India, mainly in Pollachi and the coastal region of Kerala State, produces 60% of the total world supply of white coir fibre. Sri Lanka produces 36% of the total brown fibre output. Over 50% of the coir fibre produced annually throughout the world is consumed in the countries of origin, mainly India. Together, India and Sri Lanka produce 90% of the coir produced every year.

Waste and Byproducts

Coir fibres make up about a third of the coconut pulp. The rest, called peat, pith or dust, is biodegradable, but takes 20 years to decompose. Once considered as waste material, pith is now being used as mulch, soil treatment and a hydroponic growth medium.

- Coir Board of India

- Fiber rope

- International Year of Natural Fibres 2009

- Coconut production in Kerala

Permanent Crop

A permanent crop is one produced from plants which last for many seasons, rather than being replanted after each harvest.

Traditionally, "arable land" included any land suitable for the growing of crops, even if it was actually being used for the production of permanent crops such as grapes or peaches. Modern agriculture—particularly organizations such as the CIA and FAO—prefer the term of art permanent cropland to describe such "cultivable land" that is not being used for annually-harvested crops such as staple grains. In such usage, permanent cropland is a form of "agricultural land" that includes grasslands and shrublands used to grow grape vines or coffee; orchards used to grow fruit or olives; and forested plantations used to grow nuts or rubber. It does not include, however, tree farms intended to be used for wood or timber.

Perennial Crop

Perennial crops are crops developed to reduce inputs necessary to produce food. By greatly reducing the need to replant crops from year-to-year, perennial cropping can reduce topsoil losses due to erosion, increase biological carbon sequestration within the soil, and greatly reduce waterway pollution through agricultural runoff.

Mechanisms

- Erosion Control: Because plant materials (stems, crowns, etc.) can remain in place year-round, topsoil erosion due to wind and rainfall/irrigation is reduced

- Water-use efficiency: Because these crops tend to be deeper and more fibrously-rooted than their annual counterparts, they are able to hold onto soil moisture more efficiently, while filtering pollutants (e.g. excess nitrogen) traveling to groundwater sources.

- Nutrient cycling efficiency: Because perennials more efficiently take up nutrients as a result of their extensive root systems, reduced amounts of nutrients need to be supplemented, lowering production costs while reducing possible excess sources of fertilizer runoff.

- Light interception efficiency: Earlier canopy development and longer green leaf duration increase the seasonal light interception efficiency of perennials, an important factor in plant productivity.

Example Crops

- Perennial sunflower- A perennial oil and seedcrop developed through backcrossing genes with wild sunflower.

- Perennial grain- More extensive root systems allow for more efficient water and nutrient uptake, while reducing erosion due to rain and wind year-round.

- Perennial rice- Currently in the development stage using similar methods to those used in producing the perennialized sunflower, perennial rice promises to reduce deforestation through increases in production efficiency by keeping cleared land out of the fallow stage for long periods of time.

References

- Smith, Bruce D. (1998) The Emergence of Agriculture. Scientific American Library, A Division of HPHLP, New York, ISBN 0-7167-6030-4.

- Yang, Lihui; et al. (2005). Handbook of Chinese Mythology. New York: Oxford University Press. p. 198. ISBN 978-0-19-533263-6.

- Mayrhofer, Manfred (1996). Etymologisches Wörterbuch des Altindoarischen (in German). 2. Heidelberg: Universitätsverlag Winter. p. 598. ISBN 3-8253-4550-5.

- Harris, David R. (1996). The Origins and Spread of Agriculture and Pastoralism in Eurasia. Psychology Press. p. 565. ISBN 1-85728-538-7.

- National Research Council (1996). "African Rice". Lost Crops of Africa: Volume I: Grains. Lost Crops of Africa. 1. National Academies Press. ISBN 978-0-309-04990-0. Retrieved July 18, 2008.

- Carney, Judith Ann (2001). Black rice: the African origins of rice cultivation in the Americas. Cambridge: Harvard University Press. ISBN 0-674-00452-3.

- Shahidur Rashid, Ashok Gulari and Ralph Cummings Jnr (eds) (2008) From Parastatals to Private Trade. International Food Policy Research Institute and Johns Hopkins University Press, ISBN 0-8018-8815-8

- Houtsma, M. Th.; Wensinck, A. J.; Arnold, T. W.; Heffening, W.; Lévi-Provençal, E. (eds.). "awah". First Encyclopedia of Islam. IV. E. J. Brill. p. 631. ISBN 90-04-09790-2. Retrieved January 11, 2016.

- Cousin, Tracey L. (June 1997). "Ethiopia Coffee and Trade". American University. Archived from the original on 5 November 2015. Retrieved 18 February 2016.

- Rajpal, Vijay Rani (2016). Gene Pool Diversity and Crop Improvement, Volume 1. Springer. p. 117. ISBN 978-3-319-27096-8. Retrieved 9 April 2016.

- Cole, George S. A complete Dictionary of Dry Goods and History of Silk, Linen, Wool and other Fibrous Substances. 1892. Retrieved 3 September 2015

- Gerry. "Feeding the World One Genetically Modified Tomato at a Time: A Scientific Perspective". SITN. Retrieved September 11, 2015.

- ""Wild grass became maize crop more than 8,700 years ago", National Science Foundation, News Release at Eurekalert March 24, 2009". March 23, 2009. Retrieved October 6, 2014.

Plant Pathology: An Scientific Study

The study of diseases in plants caused by pathogens and certain environmental factors is called plant pathology. This chapter examines in minutiae the common pathogens, the diseases and the infection methods used by the pathogens to spread disease in the crop. It also examines the natural and man-made causes of diseases in plants.

Plant pathology (also phytopathology) is the scientific study of diseases in plants caused by pathogens (infectious organisms) and environmental conditions (physiological factors). Organisms that cause infectious disease include fungi, oomycetes, bacteria, viruses, viroids, virus-like organisms, phytoplasmas, protozoa, nematodes and parasitic plants. Not included are ectoparasites like insects, mites, vertebrate, or other pests that affect plant health by consumption of plant tissues. Plant pathology also involves the study of pathogen identification, disease etiology, disease cycles, economic impact, plant disease epidemiology, plant disease resistance, how plant diseases affect humans and animals, pathosystem genetics, and management of plant diseases.

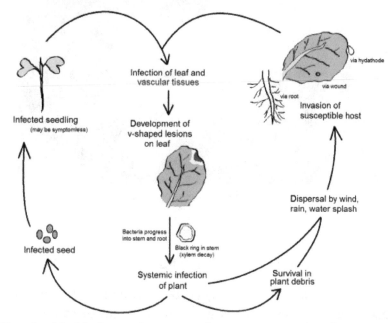

Life cycle of the black rot pathogen, *Xanthomonas campestris* pathovar *campes*

Overview

Control of plant diseases is crucial to the reliable production of food, and it provides significant reductions in agricultural use of land, water, fuel and other inputs. Plants in both natural and cultivated populations carry inherent disease resistance, but there are numerous examples of devastating plant disease impacts, as well as recurrent severe plant diseases. However, disease control is reasonably successful for most crops. Disease control is achieved by use of plants that have been bred for good resistance to many diseases, and by plant cultivation approaches such as crop rotation, use of pathogen-free seed, appropriate planting date and plant density, control of field moisture, and pesticide use. Across large regions and many crop species, it is estimated that diseases typically reduce plant yields by 10% every year in more developed settings, but yield loss to diseases often exceeds 20% in less developed settings. Continuing advances in the science of plant pathology are needed to improve disease control, and to keep up with changes in disease pressure caused by the ongoing evolution and movement of plant pathogens and by changes in agricultural practices. Plant diseases cause major economic losses for farmers worldwide. The Food and Agriculture Organization estimates indeed that pests and diseases are responsible for about 25% of crop loss. To solve this issue, new methods are needed to detect diseases and pests early, such as novel sensors that detect plant odours and spectroscopy and biophotonics that are able to diagnostic plant health and metabolism.

Plant Pathogens

Powdery mildew, a biotrophic fungus

Fungi

Most phytopathogenic fungi belong to the Ascomycetes and the Basidiomycetes.

The fungi reproduce both sexually and asexually via the production of spores and other structures. Spores may be spread long distances by air or water, or they may be soilborne. Many soil inhabiting fungi are capable of living saprotrophically, carrying out the part of their life cycle in the soil. These are known as facultative saprotrophs.

Fungal diseases may be controlled through the use of fungicides and other agriculture practices. However, new races of fungi often evolve that are resistant to various fungicides.

Biotrophic fungal pathogens colonize living plant tissue and obtain nutrients from living host cells. Necrotrophic fungal pathogens infect and kill host tissue and extract nutrients from the dead host cells. See the powdery mildew and rice blast images, below.

Rice blast, caused by a necrotrophic fungus

Significant fungal plant pathogens include:

Ascomycetes

- *Fusarium* spp. (causal agents of Fusarium wilt disease)

- *Thielaviopsis* spp. (causal agents of: canker rot, black root rot, *Thielaviopsis* root rot)

- *Verticillium* spp.

- *Magnaporthe grisea* (causal agent of rice blast)

- *Sclerotinia sclerotiorum* (causal agent of cottony rot)

Basidiomycetes

- *Ustilago* spp. (causal agents of smut)

- *Rhizoctonia* spp.

- *Phakospora pachyrhizi* (causal agent of soybean rust)

- *Puccinia* spp. (causal agents of severe rusts of virtually all cereal grains and cultivated grasses)

- *Armillaria* spp. (the so-called honey fungus species, which are virulent pathogens of trees and produce edible mushrooms)

Fungus-like Organisms

Oomycetes

The oomycetes are not true fungi but are fungus-like organisms. They include some of the most destructive plant pathogens including the genus *Phytophthora*, which includes the causal agents of potato late blight and sudden oak death. Particular species of oomycetes are responsible for root rot.

Despite not being closely related to the fungi, the oomycetes have developed very similar infection strategies. Oomycetes are capable of using effector proteins to turn off a plant's defenses in its infection process. Plant pathologists commonly group them with fungal pathogens.

Significant oomycete plant pathogens

- *Pythium* spp.

- *Phytophthora* spp., including the causal agent of the Great Irish Famine (1845–1849)

Phytomyxea

Some slime molds in Phytomyxea cause important diseases, including club root in cabbage and its relatives and powdery scab in potatoes. These are caused by species of *Plasmodiophora* and *Spongospora*, respectively.

Bacteria

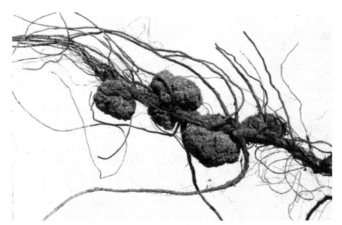

Crown gall disease caused by Agrobacterium

Most bacteria that are associated with plants are actually saprotrophic and do no harm to the plant itself. However, a small number, around 100 known species, are able to cause disease. Bacterial diseases are much more prevalent in subtropical and tropical regions of the world.

Most plant pathogenic bacteria are rod-shaped (bacilli). In order to be able to colonize the plant they have specific pathogenicity factors. Five main types of bacterial pathogenicity factors are known: uses of cell wall–degrading enzymes, toxins, effector proteins, phytohormones and exopolysaccharides.

Pathogens such as *Erwinia* species use cell wall–degrading enzymes to cause soft rot. *Agrobacterium* species change the level of auxins to cause tumours with phytohormones. Exopolysaccharides are produced by bacteria and block xylem vessels, often leading to the death of the plant.

Bacteria control the production of pathogenicity factors via quorum sensing.

Vitis vinifera with "Ca. Phytoplasma vitis" infection

Significant bacterial plant pathogens:

- Burkholderia

- Proteobacteria

    o *Xanthomonas* spp.

    o *Pseudomonas* spp.

- Pseudomonas syringae pv. tomato causes tomato plants to produce less fruit, and it "continues to adapt to the tomato by minimizing its recognition by the tomato immune system."

Phytoplasmas ('Mycoplasma-like Organisms') and Spiroplasmas

Phytoplasma and *Spiroplasma* are a genre of bacteria that lack cell walls and are related to the mycoplasmas, which are human pathogens. Together they are referred to as the mollicutes. They also tend to have smaller genomes than most other bacteria. They are normally transmitted by sap-sucking insects, being transferred into the plants phloem where it reproduces.

Tobacco mosaic virus

Viruses, Viroids and Virus-like Organisms

There are many types of plant virus, and some are even asymptomatic. Under normal circumstances, plant viruses cause only a loss of crop yield. Therefore, it is not economically viable to try to control them, the exception being when they infect perennial species, such as fruit trees.

Most plant viruses have small, single-stranded RNA genomes. However some plant viruses also have double stranded RNA or single or double stranded DNA genomes. These genomes may encode only three or four proteins: a replicase, a coat protein, a movement protein, in order to allow cell to cell movement through plasmodesmata, and sometimes a protein that allows transmission by a vector. Plant viruses can have several more proteins and employ many different molecular translation methods.

Plant viruses are generally transmitted from plant to plant by a vector, but mechanical and seed transmission also occur. Vector transmission is often by an insect (for example, aphids), but some fungi, nematodes, and protozoa have been shown to be viral vectors. In many cases, the insect and virus are specific for virus transmission such as the beet leafhopper that transmits the curly top virus causing disease in several crop plants.

Nematodes

Root-knot nematode galls

Nematodes are small, multicellular wormlike animals. Many live freely in the soil, but there are some species that parasitize plant roots. They are a problem in tropical and subtropical regions of the world, where they may infect crops. Potato cyst nematodes (*Globodera pallida* and *G. rostochiensis*) are widely distributed in Europe and North and South America and cause $300 million worth of damage in Europe every year. Root knot nematodes have quite a large host range, whereas cyst nematodes tend to be able to infect only a few species. Nematodes are able to cause radical changes in root cells in order to facilitate their lifestyle.

Protozoa and Algae

There are a few examples of plant diseases caused by protozoa (e.g., *Phytomonas*, a kinetoplastid). They are transmitted as zoospores that are very durable, and may be able to survive in a resting state in the soil for many years. They have also been shown to transmit plant viruses.

When the motile zoospores come into contact with a root hair they produce a plasmodium and invade the roots.

Some colourless parasitic algae (e.g., *Cephaleuros*) also cause plant diseases.

Parasitic Plants

Parasitic plants such as mistletoe and dodder are included in the study of phytopathology. Dodder, for example, is used as a conduit either for the transmission of viruses or virus-like agents from a host plant to a plant that is not typically a host or for an agent that is not graft-transmissible.

Common Pathogenic Infection Methods

- Cell wall-degrading enzymes: These are used to break down the plant cell wall in order to release the nutrients inside.

- Toxins: These can be non-host-specific, which damage all plants, or host-specific, which cause damage only on a host plant.

- Effector proteins: These can be secreted into the extracellular environment or directly into the host cell, often via the Type three secretion system. Some effectors are known to suppress host defense processes. This can include: reducing the plants internal signaling mechanisms or reduction of phytochemicals production. Bacteria, fungus and oomycetes are known for this function.

Physiological Plant Disorders

Natural

Drought

Frost damage and breakage by snow and hail

Flooding and poor drainage

Nutrient deficiency

Salt deposition and other soluble mineral excesses (e.g., gypsum)

Wind (windburn and breakage by hurricanes and tornadoes)

Lightning and wildfire (also often man-made)

Man-made (arguably not abiotic, but usually regarded as such)

Soil compaction

Pollution of air, soil, or both

Salt from winter road salt application or irrigation

Herbicide over-application

Poor education and training of people working with plants (e.g. lawnmower damage to trees)

Vandalism

Orchid leaves with viral infections

Epidemiology

the study of phytopathology epidemics

Management

Quarantine

A diseased patch of vegetation or individual plants can be isolated from other, healthy growth. Specimens may be destroyed or relocated into a greenhouse for treatment or study. Another option is to avoid the introduction of harmful nonnative organisms by controlling all human traffic and activity (e.g., AQIS), although legislation and enforcement are crucial in order to ensure lasting effectiveness.

Cultural

Farming in some societies is kept on a small scale, tended by peoples whose culture includes farming traditions going back to ancient times. (An example of such traditions would be lifelong training in techniques of plot terracing, weather anticipation and response, fertilization, grafting, seed care, and dedicated gardening.) Plants that are intently monitored often benefit from not only active external protection but also a greater overall vigor. While primitive in the sense of being the most labor-intensive solution by far, where practical or necessary it is more than adequate.

Plant Resistance

Sophisticated agricultural developments now allow growers to choose from among systematically cross-bred species to ensure the greatest hardiness in their crops, as suited for a particular region's pathological profile. Breeding practices have been perfected over centuries, but with the advent of genetic manipulation even finer control of a crop's immunity traits is possible. The engineering of food plants may be less rewarding, however, as higher output is frequently offset by popular suspicion and negative opinion about this "tampering" with nature.

Chemical

Many natural and synthetic compounds can be employed to combat the above threats. This method works by directly eliminating disease-causing organisms or curb-ing their spread; however, it has been shown to have too broad an effect, typically, to be good for the local eco-system. From an economic standpoint, all but the simplest natural additives may disqualify a product from "organic" status, potentially reducing the value of the yield.

Biological

Crop rotation may be an effective means to prevent a parasitic population from becoming well-es-tablished, as an organism affecting leaves would be starved when the leafy crop is replaced by a tuberous type, etc. Other means to undermine parasites without attacking them directly may exist.

Integrated

The use of two or more of these methods in combination offers a higher chance of effectiveness.

Timeline of Plant Pathology

300–286 BC Theophrastus, father of botany, wrote and studied diseases of trees, cereals and le-gumes

1665 Robert Hooke illustrates a plant-pathogenic fungal disease, rose rust

1675 Anton van Leeuwenhouek invents the compound microscope, in 1683 describes bacteria seen with the microscope

1729 Pier Antonio Micheli, father of mycology, observes spores for the first time, conducts germi-nation experiments

1755 Tillet reports on treatment of seeds

1802 Lime sulfur first used to control plant disease

1845–1849 Potato late blight epidemic in Ireland

1853 Heinrich Anton de Bary father of modern mycology, establishes that fungi are the cause, not the result, of plant diseases, publishes "Untersuchungen uber die Brandpilze"

1858 Julius Kühn publishes "Die Krankheiten der Kultergewachse"

1865 M. Planchon discovers a new species of *Phylloxera*, which was named *Phylloxera vastatrix*.

1868–1882 Coffee rust epidemic in Sri Lanka

1875 Mikhail Woronin identified the cause of clubroot as a "plasmodiophorous organism" and gave it the name *Plasmodiophora brassicae*

1876 *Fusarium oxysporum* f.sp. *cubense*, responsible for Panama disease, discovered in bananas in Australia

1878–1885 Downy mildew of grape epidemic in France

1879 Robert Koch establishes germ theory: diseases are caused by microorganisms

1882 *Lehrbuch der Baumkrankheiten* (*Textbook of Diseases of Trees*), by Robert Hartig, is published in Berlin, the first textbook of forest pathology.

1885 Bordeaux mixture introduced by Pierre-Marie-Alexis Millardet to control downy mildew on grape

1885 Experimental proof that bacteria can cause plant diseases: "Erwinia amylovora" and fire blight of apple

1886–1898 Recognition of plant viral diseases: Tobacco mosaic virus

1889 Introduction of hot water treatment of seed for disease control by Jensen

1902 First chair of plant pathology established, in Copenhagen

1904 Mendelian inheritance of cereal rust resistance demonstrated

1907 First academic department of plant pathology established, at Cornell University

1908 American Phytopathological Society founded

1910 Panama disease reaches Western Hemisphere

1911 Scientific journal *Phytopathology* founded

1925 Panama disease reaches every banana-growing country in the Western Hemisphere

1951 European and Mediterranean Plant Protection Organization (EPPO) founded

1967 Recognition of plant pathogenic mycoplasma-like organisms

1971 T. O. Diener discovers viroids, organisms smaller than viruses

The historical landmarks in plant pathology are taken from unless otherwise noted.

References

- Jackson RW (editor). (2009). Plant Pathogenic Bacteria: Genomics and Molecular Biology. Caister Academic Press. ISBN 978-1-904455-37-0.

- "Plasmopara viticola, the Cause of Downy Mildew of Grapes". The Origin of Plant Pathology and The Potato Famine, and Other Stories of Plant Diseases. Retrieved 4 February 2015.

- "Fusarium oxysporum : The End of the Banana Industry?". The Origin of Plant Pathology and The Potato Famine, and Other Stories of Plant Diseases. Retrieved 4 February 2015.

- Nicole Davis (September 9, 2009). "Genome of Irish potato famine pathogen decoded". Haas et al. Broad Institute of MIT and Harvard. Retrieved 24 July 2012.

- "Scientists discover how deadly fungal microbes enter host cells". (VBI) at Virginia Tech affiliates. Physorg. July 22, 2010. Retrieved July 31, 2012.

Importance of Crop Yielding

Crop yield refers to the seed generation by the plant and also the output per unit area of cultivated land. It is the measure that helps determine the yield potential of strains of cultivated crop which ultimately helps farmers and agricultural scientists determine the favorable conditions for higher output and also to establish better hybrid varieties of the crop. This chapter sheds light on the importance of crop yielding and gaps in yielding as well.

Crop Yield

In agriculture, crop yield (also known as "agricultural output") refers to both the measure of the yield of a crop per unit area of land cultivation, and the seed generation of the plant itself (e.g. if three grains are harvested for each grain seeded, the resulting yield is 1:3). The figure, 1:3 is considered by agronomists as the minimum required to sustain human life.

One of the three seeds must be set aside for the next planting season, the remaining two either consumed by the grower, or one for human consumption and the other for livestock feed. The higher the surplus, the more livestock can be established and maintained, thereby increasing the physical and economic well-being of the farmer and his family. This, in turn, resulted in better stamina, better overall health, and better, more efficient work. In addition, the more the surplus the more draft animals — horse and oxen — could be supported and harnessed to work, and manure, the soil thereby easing the farmer's burden. Increased crop yields meant few hands were needed on farm, freeing them for industry and commerce. This, in turn, led to the formation and growth of cities.

Formation and growth of cities meant an increased demand for food stuffs by non-farmers, and their willingness to pay for it. This, in turn, led the farmer to (further) innovation, more intensive farming, the demand/creation of new and/or improved farming implements, and a quest for improved seed which improved crop yield. Thus allowing the farmer to raise his income by bringing more food to non-farming (city) markets.

Measurement

The unit by which the yield of a crop is measured is kilograms per hectare or bushels per acre.

History

Historically speaking, a major increase in crop yield took place in the early eighteenth century with the end of the ancient, wasteful cycle of the three-course system of crop rotation whereby a third of the land lay fallow every year and hence taken out of human food, and animal feed, production.

It was to be replaced by the four-course system of crop rotation, devised in England in 1730 by Charles Townshend, 2nd Viscount Townshend or "Turnip" Townshend during the British Agricultural Revolution, as he was called by early detractors.

In the *first year* wheat or oats were planted; in the *second year* barley or oats; in the *third year* clover, rye, rutabaga and/or kale were planted; in the *fourth year* turnips were planted but not harvested. Instead, sheep were driven on to the turnip fields to eat the crop, trample the leavings under their feet into the soil, and by doing all this, fertilize the land with their droppings. In the fifth year (or *first year* of the new rotation), the cycle began once more with a planting of wheat or oats, in an average, a thirty percent increased yield.

Model of Crop yield

Crop Simulation Model

A Crop Simulation Model (CSM,SENMAP) is a simulation model that helps estimate crop yield as a function of weather conditions, soil conditions, and choice of crop management practices.

Types of Crop Simulation Models

Crop simulation models have been classified into three broad categories:

1. Statistical models: These typically rely on yield information for large areas (such as counties), and identify broad trends. The two main trends identified are a secular trend of a gradual increase in crop yield, and variation based on weather conditions. Statistical models are a significant improvement over naive historical predictions, but are not preferred for very fine-grained predictions.

2. Mechanistic models: These attempt to use fundamental mechanisms of plant and soil processes to simulate specific outcomes. These involve fairly detailed and computation-intensive simulations. These models use a continuous evolution and simulating them previously requires a small time step.

3. Functional models: These use simplified closed functional forms to simulate complex processes. They are computationally easier than mechanistic models, and can often give results that are of only somewhat worse accuracy. The Penman equation is an example of an equation that might be used as a component of a functional model. Functional models are typically run using a daily time step and the data is updated daily.

Commonly used Crop Simulation Models

- CropSyst, a multi-year multi-crop daily time-step crop simulation model developed by a team at Washington State University's Department of Biological Systems Engineering.

- APSIM, the Agricultural Production Systems sIMulator is a highly advanced simulator of agricultural systems. APSIM was created by CSIRO, the State of Queensland (through its Department of Agriculture Fisheries and Forestry) and The University of Queensland in Australia.

Principles and Practices of Crop Management

4

The main concern of crop management is to maximize agricultural output while reducing strain on the cultivated land. With this in mind a multitude of cropping methods and practices have been developed and this chapter explains in detail methods like intercropping, monocropping, sharecropping, multicropping and pollination management listing the pros and cons of each.

Intercropping

Intercropping is a multiple cropping practice involving growing two or more crops in proximity. The most common goal of intercropping is to produce a greater yield on a given piece of land by making use of resources that would otherwise not be utilized by a single crop. Careful planning is required, taking into account the soil, climate, crops, and varieties. It is particularly important not to have crops competing with each other for physical space, nutrients, water, or sunlight. Examples of intercropping strategies are planting a deep-rooted crop with a shallow-rooted crop, or planting a tall crop with a shorter crop that requires partial shade. Inga alley cropping has been proposed as an alternative to the ecological destruction of slash-and-burn farming.

When crops are carefully selected, other agronomic benefits are also achieved. Lodging-prone plants, those that are prone to tip over in wind or heavy rain, may be given structural support by their companion crop. Creepers can also benefit from structural support. Some plants are used to suppress weeds or provide nutrients. Delicate or light-sensitive plants may be given shade or protection, or otherwise wasted space can be utilized. An example is the tropical multi-tier system where coconut occupies the upper tier, banana the middle tier, and pineapple, ginger, or leguminous fodder, medicinal or aromatic plants occupy the lowest tier.

Intercropping of compatible plants also encourages biodiversity, by providing a habitat for a variety of insects and soil organisms that would not be present in a single-crop environment. This in turn can help limit outbreaks of crop pests by increasing predator biodiversity. Additionally, reducing the homogeneity of the crop increases the barriers against biological dispersal of pest organisms through the crop.

The degree of spatial and temporal overlap in the two crops can vary somewhat, but both requirements must be met for a cropping system to be an intercrop. Numerous types of intercropping, all of which vary the temporal and spatial mixture to some degree, have been identified. These are some of the more significant types:

- Mixed intercropping, as the name implies, is the most basic form in which the component crops are totally mixed in the available space.

- Row cropping involves the component crops arranged in alternate rows. Variations include alley cropping, where crops are grown in between rows of trees, and strip cropping, where multiple rows, or a strip, of one crop are alternated with multiple rows of another crop. A new version of this is to intercrop rows of solar photovoltaic modules with agriculture crops. This practice is called agrivoltaics.

- Temporal intercropping uses the practice of sowing a fast-growing crop with a slow-growing crop, so that the fast-growing crop is harvested before the slow-growing crop starts to mature.

- Further temporal separation is found in relay cropping, where the second crop is sown during the growth, often near the onset of reproductive development or fruiting, of the first crop, so that the first crop is harvested to make room for the full development of the second.

Chili pepper intercropped with coffee in Colombia's southwestern Cauca Department	Coconut and *Tagetes erecta*, a multilayer cropping in India

Multiple Cropping

In agriculture, multiple cropping is the practice of growing two or more crops in the same piece of land during a single growing season. It is a form of polyculture. It can take the form of double-cropping, in which a second crop is planted after the first has been harvested, or relay cropping, in which the second crop is started amidst the first crop before it has been harvested. A related practice, companion planting, is sometimes used in gardening and intensive cultivation of vegetables and fruits. One example of multi-cropping is tomatoes + onions + marigold; the marigolds repel some tomato pests.

Multiple cropping is found in many agricultural traditions. In the Garhwal Himalaya of India, a practice called baranaja involves sowing 12 or more crops on the same plot, including various types of beans, grains, and millets, and harvesting them at different times.

In the cultivation of rice, multiple cropping requires effective irrigation, especially in areas with a dry season. Rain that falls during the wet season permits the cultivation of rice during that period, but during the other half of the year, water cannot be channeled into the rice fields without an irrigation system. The Green Revolution in Asia led to the development of high-yield varieties of rice, which required a substantially shorter growing season of 100 days, as opposed to traditional varieties, which needed 150 to 180 days. Due to this, multiple cropping became more prevalent in Asian countries.

Monocropping

Monocropping is the agricultural practice of growing a single crop year after year on the same land, in the absence of rotation through other crops or growing multiple crops on the same land (polyculture). Corn, soybeans, and wheat are three common crops often grown using monocropping techniques.

While economically a very efficient system, allowing for specialization in equipment and crop production, monocropping is also controversial, as it can damage the soil ecology (including depletion or reduction in diversity of soil nutrients) and provide an unbuffered niche for parasitic species, increasing crop vulnerability to opportunistic insects, plants, and microorganisms. The result is a more fragile ecosystem with an increased dependency on pesticides and artificial fertilizers. The concentrated presence of a single cultivar, genetically adapted with a single resistance strategy, presents a situation in which an entire crop can be wiped out very quickly by a single opportunistic species. An example of this would be the potato famine of Ireland in 1845–1849, and according to Devlin Kuyek is the main cause of the current food crisis with monoculture rice crops failing as the effects of climate change become more acute.

Monocropping as an agricultural strategy tends to emphasize the use of expensive specialized farm equipment — an important component in realizing its efficiency goals. This can lead to an increased dependency on fossil fuels and reliance on expensive machinery that cannot be produced locally and may need to be financed. This can make a significant change in the economics of farming in regions that are accustomed to self-sufficiency in agricultural production. In addition, political complications may ensue when these dependencies extend across national boundaries.

The controversies surrounding monocropping are complex, but traditionally the core issues concern the balance between its advantages in increasing short-term food production — especially in hunger-prone regions — and its disadvantages with respect to long-term land stewardship and the fostering of local economic independence and ecological sustainability. Advocates of monocropping tend to claim that in its absence many human populations would be reduced to starvation or to a degraded level of civilization comparable to the Dark Ages. On the other hand, critics of monocropping dispute these claims and attribute them to corporate special interest groups, citing the damage that monocropping causes to societies and the environment.

A difficulty with monocropping is that the solution to one problem — whether economic, environmental or political — may result in a cascade of other problems. For example, a well-known concern is pesticides and fertilizers seeping into surrounding soil and groundwater from extensive monocropped acreage in the U.S. and abroad. This issue, especially with respect to the pesticide DDT, played an important role in focusing public attention on ecology and pollution issues during the 1960s when Rachel Carson published her landmark book Silent Spring.

Soil depletion is also a negative effect of mono-cropping. Crop rotation plays an important role in replenishing soil nutrients, especially atmospheric nitrogen converted to usable forms by nitrogen-fixing plants used in fallow fields. In addition, it performs an important role in preventing pathogen and pest build-up. In a monocropping regime, farmers are less likely to rotate their crops and replenish such essential soil nutrients. In addition, artificial high-nitrogen fertilizers can "burn" the soil by creating an unfavorable environment for indigenous organisms, a phenomenon well-known to organic gardeners and farmers (who avoid it), resulting in further disruption of soil ecology and dependence on further short-term fertilizer strategies. Lacking a stable ecology, in the absence of substantial irrigation and chemical "fixes" the soil can become dry and begin to erode. As the soil becomes arid and useless, the need for more land becomes an issue, leading to the destruction of even more land — a high-tech version of slash and burn agriculture.

Under certain circumstances monocropping can lead to deforestation (Tauli-Corpuz;Tamang, 2007) or the displacement of indigenous peoples (Tauli-Corpuz;Tamang, 2007).

In order to help reduce dependence on fossil fuels the U.S. government subsidizes the monocropping of corn and soybeans to be used in ethanol production (S, 2007). However monocropping itself is highly chemical- and energy-intensive, as studies by Nelson (2006) indicate. Such studies have shown that the "hidden" energy costs associated with producing each unit of bio-fuel are significantly larger than the amount of energy available from the fuel itself.

Sharecropping

A FSA photo of a cropper family chopping the weeds from cotton near White Plains, in Georgia, USA (1941).

Sharecropping is a system of agriculture in which a landowner allows a tenant to use the land in return for a share of the crops produced on their portion of land. Sharecropping has a long history and there are a wide range of different situations and types of agreements that have used a form of the system. Some are governed by tradition, and others by law. Legal contract systems such as the Italian *mezzadria*, the French *métayage*, the Spanish *mediero*, or the Islamic system of *muqasat*, occur widely.

Overview

Sharecropping has benefits and costs for both the owners and the tenant. It encourages the cropper to remain on the land, solving the harvest rush problem. At the same time, since the cropper pays in shares of his harvest, owners and croppers share the risks of harvests being large or small and of prices being high or low. Because tenants benefit from larger harvests, they have an incentive to work harder and invest in better methods than in a slave plantation system. However, by dividing the working force into many individual workers, large farms no longer benefit from economies of scale. On the whole, sharecropping was not as economically efficient as the gang agriculture of slave plantations.

In the U.S. "tenant" farmers own their own mules and equipment, and "sharecroppers" do not, and thus sharecroppers are poorer and of lower status. Sharecropping occurred extensively in Scotland, Ireland and colonial Africa, and came into wide use in the Southern United States during the Reconstruction era (1865–1877). The South had been devastated by war - planters had ample land but little money for wages or taxes. At the same time, most of the former slaves had labor but no money and no land - they rejected the kind of gang labor that typified slavery. A solution was the sharecropping system focused on cotton, which was the only crop that could generate cash for the croppers, landowners, merchants and the tax collector. Poor white farmers, who previously had done little cotton farming, needed cash as well and became sharecroppers.

Jeffery Paige made a distinction between centralized sharecropping found on cotton plantations and the decentralized sharecropping with other crops. The former is characterized by political conservatism and long lasting tenure. Tenants are tied to the landlord through the plantation store. Their work is heavily supervised as slave plantations were. This form of tenure tends to be replaced by wage slavery as markets penetrate. Decentralized sharecropping involves virtually no role for the landlord: plots are scattered, peasants manage their own labor and the landowners do not manufacture the crops. Leases are very short which leads to peasant radicalism. This form of tenure becomes more common when markets penetrate.

Use of the sharecropper system has also been identified in England (as the practice of "farming to halves"). It is still used in many rural poor areas today, notably in Pakistan and India.

Although there is a perception that sharecropping was exploitative, "evidence from around the world suggests that sharecropping is often a way for differently endowed enterprises to pool resources to mutual benefit, overcoming credit restraints and helping to manage risk."

It can have more than a passing similarity to serfdom or indenture, particularly where associated with large debts at a plantation store that effectively tie down the workers and their family to the land. It has therefore been seen as an issue of land reform in contexts such as the Mexican Revolution. However, Nyambara states that Eurocentric historiographical devices such as 'feudalism' or 'slavery' often qualified by weak prefixes like 'semi-' or 'quasi-' are not helpful in understanding the antecedents and functions of sharecropping in Africa.

Sharecropping agreements can however be made fairly, as a form of tenant farming or sharefarming that has a variable rental payment, paid in arrears. There are three different types of contracts.

1. Workers can rent plots of land from the owner for a certain sum and keep the whole crop.

2. Workers work on the land and earn a fixed wage from the land owner but keep some of the crop.

3. No money changes hands but the worker and land owner each keep a share of the crop.

Advantages

The advantages of sharecropping in other situations include enabling access for women to arable land where ownership rights are vested only in men.

Maggie pointed out that sharecropping was economically inefficient in a free market. However, many outside factors make it efficient. One factor is slave emancipation: sharecropping provided the freed slaves of the USA, Brazil and the late Roman Empire with land access. It is efficient also as a way of escaping inflation, hence its rise in 16th century France and Italy.

Landlords opt for sharecropping to avoid the administrative costs and shirking that occurs on plantations and haciendas. It is preferred to cash tenancy because cash tenants take all the risks, and any harvest failure will hurt them and not the landlord. Therefore, they tend to demand lower rents than sharecroppers.

Disadvantages

In the southern United States, the disadvantages of sharecropping soon became apparent. Most sharecroppers and landowners were Baptist, so that ministers had access to candid opinions across the South. They reported that the system was wasteful for the landowners who dealt with tenants who were poor managers of their time and effort, and often got deeper and deeper into debt. It was also harmful to tenants for it undermined their ambition and encouraged slothfulness. It left the cropper vulnerable to intimidation and shortchanging. Nevertheless, it appeared to be inevitable, with no serious alternative unless the croppers left agriculture.

A new system of credit, the crop lien, became closely associated with sharecropping. Under this system, a planter or merchant extended a line of credit to the sharecropper while taking the year's crop as collateral. The sharecropper could then draw food and supplies all year long. When the crop was harvested, the planter or merchants who held the lien sold the harvest for the sharecropper and settled the debt.

Regions

Africa

In settler colonies of colonial Africa, sharecropping was a feature of the agricultural life. White farmers, who owned most of the land, were frequently unable to work the whole of their farm for lack of capital. They therefore allowed African farmers to work the excess on a sharecropping basis. In South Africa the 1913 Natives' Land Act outlawed the ownership of land by Africans in areas designated for white ownership and effectively reduced the status of most sharecroppers to tenant farmers and then to farm laborers. In the 1960s, generous subsidies to white farmers meant that most farmers could afford to work their entire farms, and sharecropping faded out.

The arrangement has reappeared in other African countries in modern times, including Ghana and Zimbabwe.

United States

Sharecropping became widespread in the South as a response to economic upheaval caused by the end of slavery during and after Reconstruction. Sharecropping was a way for very poor farmers, both White and Black, to earn a living from land owned by someone else. The landowner provided land, housing, tools and seed, and perhaps a mule, and a local merchant provided food and supplies on credit. At harvest time the sharecropper received a share of the crop (from one-third to one-half, with the landowner taking the rest). The cropper used his share to pay off his debt to the merchant.

Sharecroppers on the roadside after eviction (1936).

The system started with Blacks when large plantations were subdivided. By the 1880s white farmers also became sharecroppers. The system was distinct from that of the tenant farmer, who rented the land, provided his own tools and mule, and received half the crop. Landowners provided more supervision to sharecroppers, and less or none to tenant farmers. Sharecropping in the United States probably originated in the Natchez District, roughly centered in Adams County, Mississippi with its county seat, Natchez.

Sharecroppers worked a section of the plantation independently, usually growing cotton, tobacco, rice, sugar and other cash crops and received half of the parcel's output.

Although the sharecropping system was primarily a post-Civil War development, it did exist in antebellum Mississippi, especially in the northeastern part of the state, an area with few slaves or plantations, and most likely existed in Tennessee. Sharecropping, along with tenant farming, was a dominant form in the cotton South from the 1870s to the 1950s, among both blacks and whites.

An early 20th century Texas sharecropper's home diorama at the Audie Murphy American Cotton Museum, 2015.

Following the Civil War of the United States, the South lay in ruins. Plantations and other lands throughout the Southern United States were seized by the federal government and thousands of freed Black slaves known as freedmen, found themselves free, yet without means to support their families. The situation was made more complex due to General William T. Sherman's Special Field Order Number 15, which in January 1865, announced he would temporarily grant newly freed families 40 acres of land on the islands and coastal regions of Georgia. This policy was also referred to as Forty Acres and a Mule. Many believed that this policy would be extended to all former slaves and their families as repayment for their treatment at the end of the war.

An alternative path was selected and enforced. Three months later in the summer of 1865, President Andrew Johnson, as one of the first acts of Reconstruction, instead ordered all land under federal control be returned to its previous owners. This meant that plantation and land owners in the South regained their land but lacked a labor force. The solution was to use Sharecropping. It would allow the government to match labor with demand and begin the process of economically rebuilding the nation via labor contracts.

In Reconstruction-era United States, sharecropping was one of few options for penniless freedmen to conduct subsistence farming and support themselves and their families. Other solutions included the crop-lien system (where the farmer was extended credit for seed and other supplies by the merchant), a rent labor system (where the former slave rents their land but keeps their entire crop), and the wage system (worker earns a fixed wage, but keeps none of their crop). Sharecropping was by far the most economically efficient, as it provided incentives for workers to produce a bigger harvest. It was a stage beyond simple hired labor, because the sharecropper had an annual contract. During Reconstruction, the federal Freedmen's Bureau ordered the arrangements and wrote and enforced the contracts.

After the Civil War, plantation owners had to borrow money to produce crops. Interest rates on these loans were around 15%. The indebtedness of cotton planters increased through the early 1940s, and the average plantation fell into bankruptcy about every twenty years. It is against this backdrop that the wealthiest owners maintained their concentrated ownership of the land.

Cotton sharecroppers, Hale County, Alabama, 1936

A sharecropper family in Walker County, Alabama (c. 1937)

Sharecropper's cabin displayed at Louisiana State Cotton Museum in Lake Providence, Louisiana (2013 photo)

Croppers were assigned a plot of land to work, and in exchange owed the owner a share of the crop at the end of the season, usually one-half. The owner provided the tools and farm animals. Farmers who owned their own mule and plow were at a higher stage and are called tenant farmers; they paid the landowner less, usually only a third of each crop. In both cases the farmer kept the produce of gardens.

Sharecroppers' chapel at Cotton Museum in Lake Providence

The sharecropper purchased seed, tools and fertilizer, as well as food and clothing, on credit from a local merchant, or sometimes from a plantation store. When the harvest came, the cropper would harvest the whole crop and sell it to the merchant who had extended credit. Purchases and the landowner's share were deducted and the cropper kept the difference—or added to his debt.

Inside living room/bedroom combination of sharecroppers in Lake Providence

Though the arrangement protected sharecroppers from the negative effects of a bad crop, many sharecroppers (both black and white) remained quite poor. Arrangements typically gave a third of the crop to the sharecropper.

By the early 1930s there were 5.5 million white tenants, sharecroppers, and mixed cropping/laborers in the United States, and 3 million blacks. In Tennessee whites made up two thirds or more of the sharecroppers. In Mississippi, by 1900, 36% of all white farmers were tenants or sharecroppers, while 85 percent of black farmers were. Sharecropping continued to be a significant institution in Tennessee agriculture for more than sixty years after the Civil War, peaking in importance in the early 1930s, when sharecroppers operated approximately one-third of all farm units in the state.

The commissary or company store for sharecroppers at Lake Providence as it appeared in the 19th century

The situation of landless farmers who challenged the system in the rural south as late as 1941 has been described thus: "he is at once a target subject of ridicule and vitriolic denunciation; he may even be waylaid by hooded or unhooded leaders of the community, some of whom may be public officials. If a white man persists in 'causing trouble', the night riders may pay him a visit, or the officials may haul him into court; if he is a Negro, a mob may hunt him down."

Sharecroppers formed unions in the 1930s, beginning in Tallapoosa County, Alabama in 1931, and Arkansas in 1934. Membership in the Southern Tenant Farmers Union included both blacks and poor whites. As leadership strengthened, meetings became more successful, and protest became more vigorous, landlords responded with a wave of terror.

Sharecroppers' strikes in Arkansas and the Bootheel of Missouri, the 1939 Missouri Sharecroppers' Strike, were documented in the film *Oh Freedom After While*. The plight of a sharecropper was addressed in the song *Sharecropper's Blues* recorded by Charlie Barnet and His Orchestra with vocals by Kay Starr (Decca 24264) in 1944. It was rerecorded and released by Capitol with Starr being backed by the "Dave Cavanaugh Ork" (Capitol Americanna 40051). Decca then reissued the Barnet/Star recording.

In the 1930s and 1940s, increasing mechanization virtually brought the institution of sharecropping to an end in the United States. The sharecropping system in the U.S. increased during the Great Depression with the creation of tenant farmers following the failure of many small farms throughout the Dust bowl. Traditional sharecropping declined after mechanization of farm work became economical in the mid-20th century. As a result, many sharecroppers were forced off the farms, and migrated to cities to work in factories, or become migrant workers in the Western United States during World War II.

Sharecropping Agreements

Typically, a sharecropping agreement would specify which party was expected to cover certain expenses, like seed, fertilizer, weed control, irrigation district assessments, and fuel. Sometimes the sharecropper covered those costs, but they expected a larger share of the crop in return. The agreement would also indicate whether the sharecropper would use his own equipment to raise the crops, or use the landlord's equipment. The agreement would also indicate whether the landlord would pick up his or her share of the crop in the field or whether the sharecropper would deliver it (and where it would be delivered.)

For example, a landowner may have a sharecropper farming an irrigated hayfield. The sharecropper uses his own equipment, and covers all the costs of fuel and fertilizer. The landowner pays the irrigation district assessments and does the irrigating himself. The sharecropper cuts and bales the hay, and delivers one-third of the baled hay to the landlord's feedlot. The sharecropper might also leave the landlord's share of the baled hay in the field, where the landlord would fetch it when he wanted hay.

Another arrangement could have the sharecropper delivering the landlord's share of the product to market, in which case the landlord would get his share in the form of the sale proceeds. In that case, the agreement should indicate the timing of the delivery to market, which can have a significant effect on the ultimate price of some crops. The market timing decision should probably be decided shortly before harvest, so that the landlord has more complete information about the area's harvest, to determine whether the crop will earn more money immediately after harvest, or whether it should be stored until the price rises. Market timing can entail storage costs and losses to spoilage as well, for some crops.

Farmer's Cooperatives

Cooperative farming exists in many forms throughout the United States, Canada, and the rest of the world. Various arrangements can be made through collective bargaining or purchasing to get the best deals on seeds, supplies, and equipment. For example, members of a farmers' cooperative who cannot afford heavy equipment of their own can lease them for nominal fees from the cooperative. Farmers' cooperatives can also allow groups of small farmers and dairymen to manage pricing and prevent undercutting by competitors.

Economic Theories of Share Tenancy

The theory of share tenancy was long dominated by Alfred Marshall's famous footnote in Book VI, Chapter X.14 of *Principles* where he illustrated the inefficiency of agricultural share-contracting. Steven N.S. Cheung (1969), challenged this view, showing that with sufficient competition and in the absence of transaction costs, share tenancy will be equivalent to competitive labor markets and therefore efficient.

He also showed that in the presence of transaction costs, share-contracting may be preferred to either wage contracts or rent contracts—due to the mitigation of labor shirking and the provision of risk sharing. Joseph Stiglitz (1974, 1988), suggested that if share tenancy is only a labor contract, then it is only pairwise-efficient and that land-to-the-tiller reform would improve social efficiency by removing the necessity for labor contracts in the first place.

Reid (1973), Murrel (1983), Roumasset (1995) and Allen and Lueck (2004) provided transaction cost theories of share-contracting, wherein tenancy is more of a partnership than a labor contract and both landlord and tenant provide multiple inputs. It has been also argued that the sharecropping institution can be explained by factors such as informational asymmetry (Hallagan, 1978;

Allen, 1982; Muthoo, 1998), moral hazard (Reid, 1976; Eswaran and Kotwal, 1985; Ghatak and Pandey, 2000), intertemporal discounting (Roy and Serfes, 2001), price fluctuations (Sen, 2011) or limited liability (Shetty, 1988; Basu, 1992; Sengupta, 1997; Ray and Singh, 2001).

Pollination Management

Pollination management is the label for horticultural practices that accomplish or enhance pollination of a crop, to improve yield or quality, by understanding of the particular crop's pollination needs, and by knowledgeable management of pollenizers, pollinators, and pollination conditions.

Honey Bee on Domestic Plum Blossom

Honey bees are especially well adapted to collecting and moving pollen, thus are the most commonly used crop pollinators. Note the light brown pollen in the pollen basket.

Placing honey bees for pumpkin pollination
Mohawk Valley, NY

Date pollinator up an 'Abid Rahim' palm tree

Pollinator Decline

With the decline of both wild and domestic pollinator populations, pollination management is becoming an increasingly important part of horticulture. Factors that cause the loss of pollinators include pesticide misuse, unprofitability of beekeeping for honey, rapid transfer of pests and diseases to new areas of the globe, urban/suburban development, changing crop patterns, clearcut logging (particularly when mixed forests are replaced by monoculture pine), clearing of hedgerows and other wild areas, bad diet because of loss of floral biodiversity, and a loss of nectar corridors for migratory pollinators.

Importance

The increasing size of fields and orchards (monoculture) increase the importance of pollination management. Monoculture can cause a brief period when pollinators have more food resources than they can use (but monofloral diet can reduce their immune system) while other periods of the year can bring starvation or pesticide contamination of food sources. Most nectar source and pollen source throughout the growing season to build up their numbers.

Crops that traditionally have had managed pollination include apple, almonds, pears, some plum and cherry varieties, blueberries, cranberries, cucumbers, cantaloupe, watermelon, alfalfa seeds, onion seeds, and many others. Some crops that have traditionally depended entirely on chance pollination by wild pollinators need pollination management nowadays to make a profitable crop.

Some crops, especially when planted in a monoculture situation, require a very high level of pollinators to produce economically viable crops. This may be because of lack of attractiveness of the blossoms, or from trying to pollinate with an alternative when the native pollinator is extinct or rare. These include crops such as alfalfa, cranberries, and kiwifruit. This technique is known as saturation pollination. In many such cases, various native bees are vastly more efficient at pollination (e.g., with blueberries), but the inefficiency of the honey bees is compensated for by using large numbers of hives, the total number of foragers thereby far exceeding the local abundance of native pollinators. In a very few cases, it has been possible to develop commercially viable pollination techniques that use the more efficient pollinators, rather than continued reliance on honey bees, as in the management of the alfalfa leafcutter bee.

Number of hives needed per unit area of crop pollination			
Common name	**number of hives per acre**	**number of hives per hectare**	**number of bee visits per square meter/minute**
Alfalfa	1, (3–5)	2.5, (4.9–12)	
Almonds	2–3	4.9–7.4	
Apples (normal size)	1	2.5	
Apples (semi dwarf)	2	4.9	
Apples (dwarf)	3	7.4	
Apricots	1	2.5	
Blueberries	3–4	7.4–9.9	2.5
Borage	0.6–1.0	1.5–2.5	
Buckwheat	0.5–1	1.2–2.5	
Canola	1	2.5	
Canola (hybrid)	2.0–2.5	4.9–6.2	
Cantaloupes	2–4, (average 2.4)	4.9–9.9, (average 5.9)	
Clovers	1–2	2.5–4.9	
Cranberries	3	7.4	
Cucumbers	1–2, (average 2.1)	2.5–4.9, (average 5.2)	
Ginseng	1	2.5	
Muskmelon	1–3	2.5–7.4	
Nectarines	1	2.5	
Peaches	1	2.5	
Pears	1	2.5	
Plums	1	2.5	
Pumpkins	1	2.5	
Raspberries	0.7–1.3	1.7–3.2	
Squash	1–3	2.5–7.4	
Strawberries	1–3.5	2.5–8.6	
Sunflower	1	2.5	
Trefoil	0.6–1.5	1.5–3.7	
Watermelon	1–3, (average 1.3)	2.5–4.9, (average 3.2)	
Zucchini	1	2.5	

It is estimated that about one hive per acre will sufficiently pollinate watermelons. In the 1950s when the woods were full of wild bee trees, and beehives were normally kept on most South Carolina farms, a farmer who grew ten acres (4 ha) of watermelons would be a large grower and probably had all the pollination needed. But today's grower may grow 200 acres (80 ha), and, if lucky, there might be one bee tree left within range. The only option in the current economy is to bring beehives to the field during blossom time.

Types of Pollinators

Organisms that are currently being used as pollinators in managed pollination are honey bees, bumblebees, alfalfa leafcutter bees, and orchard mason bees. Other species are expected to be added to this list as this field develops. Humans also can be pollinators, as the gardener who hand pollinates her squash blossoms, or the Middle Eastern farmer, who climbs his date palms to pollinate them.

The Cooperative extension service recommends one honey bee hive per acre (2.5 hives per hectare) for standard watermelon varieties to meet this crop's pollination needs. In the past, when fields were small, pollination was accomplished by a mix of bees kept on farms, bumblebees, carpenter bees, feral honey bees in hollow trees and other insects. Today, with melons planted in large tracts, the grower may no longer have hives on the farm; he may have poisoned many of the pollinators by spraying blooming cotton; he may have logged off the woods, removing hollow trees that provided homes for bees, and pushed out the hedgerows that were home for solitary native bees and other pollinating insects.

Planning for Improved Pollination

US migratory commercial beekeeper moving spring bees from South Carolina to Maine for blueberry pollination

Before pollination needs were understood, orchardists often planted entire blocks of apples of a single variety. Because apples are self-sterile, and different members of a single variety are genetic clones (equivalent to a single plant), this is not a good idea. Growers now supply pollenizers, by planting crab apples interspersed in the rows, or by grafting crab apple limbs on some trees. Pollenizers can also be supplied by putting drum bouquets of crab apples or a compatible apple variety in the orchard blocks.

The field of pollination management cannot be placed wholly within any other field, because it bridges several fields. It draws from horticulture, apiculture, zoology (especially entomology), ecology, and botany.

Improving Pollination with Suboptimal Bee Densities

Growers' demand for beehives far exceeds the available supply. The number of managed beehives in the US has steadily declined from close to 6 million after WWII, to less than 2.5 million today. In contrast, the area dedicated to growing bee-pollinated crops has grown over 300% in the same time period. To make matters worse, in the past five years we have seen a decline in winter managed beehives, which has reached an unprecedented rate near 30%. At present, there is an enormous demand for beehive rentals that cannot always be met. There is a clear need across the agricultural industry for a management tool to draw pollinators into cultivations and encourage them to preferentially visit and pollinate the flowering crop. By attracting pollinators like honeybees and increasing their foraging behavior, particularly in the center of large plots, we can increase grower returns and optimize yield from their plantings.

Soybean Management Practices

Soybean management practices in farming are the decisions a producer must make in order to raise a soybean crop. The type of tillage, plant population, row spacing, and planting date are four major management decisions that soybean farmers must consider. How individual producers choose to handle each management application depends on their own farming circumstances.

Tillage

Tillage is defined in farming as the disturbance of soil in preparation for a crop. Tillage is usually done to warm the seed bed up, to help dry the seed bed out, to break up disease, and to reduce early weed pressure. Tillage prior to planting adds cost per acre farmed and has not been shown to increase soybean yields significantly.

"No till" is the practice of planting seed directly into the ground without disturbing the soil prior to planting. This practice eliminates the tillage pass, saving the farmer the associated costs. Planting no-till places seed directly into a cooler and wetter seed bed, which can slow germination. This process is considered a good conservation practice because tilling disturbs the soil crust, causing erosion. The practice of no-till is currently on the rise among farmers in the midwestern United States.

Plant Population

Plant population is the number of seeds planted in a given area. Population is usually expressed as plants or seeds per acre. Plant population is one of two major factors that determine canopy closure (when the plants cover the space in between the rows) and other yield components. A higher seed population will close the canopy faster and reduce soil moisture loss. A high plant population does not necessarily equal a high yield. The recommended seeding rate is 125,000 to 140,000 seeds per acre. The goal is to achieve a final stand of 100,000 plants per acre. Planting the extra seed gives the producer added insurance that each acre will attain a final stand of 100,000 plants.

Row Spacing

Row spacing is the second factor that determines canopy closure and yield components. Row spacing can either refer to the space between plants in the same row or the distance between two rows. Row spacing determines the degree of plant to plant competition. Rows planted closer together in a population will decrease the space between plants. Closer row widths increase plant to plant competition for nutrients and water but also increase sunlight use efficiency. According to former Iowa State University Soybean Extension Specialist Palle Pedersen, current recommendations are to plant rows that are less than 30" apart. This increases light interception and decreases weed competition.

Planting Date

Planting date refers to the date that the seed is sown. This concept is of prime importance in growing soybeans because yields are strongly correlated with planting date. Data from Iowa State University shows that earlier planted soybeans tend to have higher yields than soybeans planted later in the growing season. Producers seeding early need to check that the seedbed is in the right conditions (temperature, moisture, nutrients) since planting into a sub-optimal seedbed will lose yield instead of gaining it. Other special considerations include soil pathogens, insect pressure, and the possibility of frost. Fungicide treatment and insecticide treatment are options used to help reduce the risk of soil pathogens and insect pressure. Knowing the chance of frost in a given area prior to planting helps determine how early seed can safely be sown.

References

* Woodman, Harold D. (1995). New South – New Law: The legal foundations of credit and labor relations in the Postbellum agricultural South. Louisiana State University Press. ISBN 0-8071-1941-5.

* The Devil's Music: A History of the Blues By Giles Oakley Edition: 2. Da Capo Press, 1997, page 184. ISBN 0-306-80743-2, ISBN 978-0-306-80743-5

* Allen, Douglas W.; Dean Lueck (2004). The Nature of the Farm: Contracts, Risk, and Organization in Agriculture. MIT Press. p. 258. ISBN 9780262511858.

* Frank J. Dainello & Roland Roberts. "Cultural Practices". Texas Vegetable Grower's Handbook. Texas Agricultural Extension Service. Retrieved 9 December 2014.

* Griffiths, Liz Farming to Halves: A New Perspective on an Absurd and Miserable System in Rural History Today, Issue 6:2004 p.5, accessed at British Agricultural History Society, 16 February 2013.

* Roy, Jaideep; Konstantinos Serfes (2001). "Intertemporal discounting and tenurial contracts,". Journal of Development Economics. 64 (2): 417–436. doi:10.1016/S0304-3878(00)00144-9. Retrieved 2012-09-04.

* Sen, Debapriya (2011). "A theory of sharecropping: the role of price behavior and imperfect competition,". Journal of Economic Behavior & Organization. 80 (1): 181–199. doi:10.1016/j.jebo.2011.03.006. Retrieved 2012-09-04.

* Ghatak, Maitreesh; Priyanka Pandey (2000). "Contract choice in agriculture with joint moral hazard in effort and risk,". Journal of Development Economics. 63 (2): 303–326. doi:10.1016/S0304-3878(00)00116-4. Retrieved 2011-04-22.

* Shetty, Sudhir (1988). "Limited liability, wealth differences, and the tenancy ladder in agrarian economies,". Journal of Development Economics. 29: 1–22. doi:10.1016/0304-3878(88)90068-5. Retrieved 2011-04-22.

- Basu, Kaushik (1992). "Limited liability and the existence of share tenancy,". Journal of Development Economics. 38: 203–220. doi:10.1016/0304-3878(92)90026-6. Retrieved 2011-04-22.

- Sengupta, Kunal (1997). "Limited liability, moral hazard and share tenancy,". Journal of Development Economics. 52 (2): 393–407. doi:10.1016/S0304-3878(96)00444-0. Retrieved 2011-04-22.

- Ray, Tridip; Nirvikar Singh (2001). "Limited liability, contractual choice and the tenancy ladder,". Journal of Development Economics. 66: 289–303. doi:10.1016/S0304-3878(01)00163-8. Retrieved 2011-04-22.

- Muthoo, Abhinay (1998). "Renegotiation-proof tenurial contracts as screening mechanisms,". Journal of Development Economics. 56: 1–26. doi:10.1016/S0304-3878(98)00050-9. Retrieved 2011-04-22.

Crop Protection and its Techniques

Apart from hybridization of crops, another obvious method of increasing crop output is by the use of pesticides and insecticides to minimize the loss in crop that occurs due to pests and insects. This chapter elucidates the various methods employed for this purpose- spraying of pesticides and insecticides, bird netting, biological alternative to manmade pesticides and the like. Each of these methods has been explained in detail citing examples wherever needed.

Crop Protection

Crop protection is the science and practice of managing plant diseases, weeds and other pests (both vertebrate and invertebrate)that damage agricultural crops and forestry. Agricultural crops include field crops (maize, wheat, rice, etc.), vegetable crops (potatoes, cabbages, etc.) and fruits. The crops in field are exposed to many factor. The crop plants may be damaged by insects, birds, rodents, bacteria, etc. Crop protection encompasses:

- Pesticide-based approaches such as herbicides, insecticides and fungicides

- Biological pest control approaches such as cover crops, trap crops and beetle banks

- Barrier-based approaches such as agrotextiles and bird netting

- Animal psychology-based approaches such as bird scarers

- Biotechnology-based approaches such as plant breeding and genetic modification

Crop Protection Encompasses

Pesticide

A Lite-Trac four-wheeled self-propelled crop sprayer spraying pesticide on a field

Pesticides are substances meant for attracting, seducing, and then destroying any pest. They are a class of biocide. The most common use of pesticides is as plant protection products (also known as crop protection products), which in general protect plants from damaging influences such as weeds, fungi, or insects. This use of pesticides is so common that the term *pesticide* is often treated as synonymous with *plant protection product*, although it is in fact a broader term, as pesticides are also used for non-agricultural purposes. The term pesticide includes all of the following: herbicide, insecticide, insect growth regulator, nematicide, termiticide, molluscicide, piscicide, avicide, rodenticide, predacide, bactericide, insect repellent, animal repellent, antimicrobial, fungicide, disinfectant (antimicrobial), and sanitizer.

A crop-duster spraying pesticide on a field

In general, a pesticide is a chemical or biological agent (such as a virus, bacterium, antimicrobial, or disinfectant) that deters, incapacitates, kills, or otherwise discourages pests. Target pests can include insects, plant pathogens, weeds, mollusks, birds, mammals, fish, nematodes (roundworms), and microbes that destroy property, cause nuisance, or spread disease, or are disease vectors. Although pesticides have benefits, some also have drawbacks, such as potential toxicity to humans and other species. According to the Stockholm Convention on Persistent Organic Pollutants, 9 of the 12 most dangerous and persistent organic chemicals are organochlorine pesticides.

Definition

Type of pesticide	Target pest group
Herbicides	Plant
Algicides or Algaecides	Algae
Avicides	Birds

Bactericides	Bacteria
Fungicides	Fungi and Oomycetes
Insecticides	Insects
Miticides or Acaricides	Mites
Molluscicides	Snails
Nematicides	Nematodes
Rodenticides	Rodents
Virucides	Viruses

The Food and Agriculture Organization (FAO) has defined *pesticide* as:

> any substance or mixture of substances intended for preventing, destroying, or controlling any pest, including vectors of human or animal disease, unwanted species of plants or animals, causing harm during or otherwise interfering with the production, processing, storage, transport, or marketing of food, agricultural commodities, wood and wood products or animal feedstuffs, or substances that may be administered to animals for the control of insects, arachnids, or other pests in or on their bodies. The term includes substances intended for use as a plant growth regulator, defoliant, desiccant, or agent for thinning fruit or preventing the premature fall of fruit. Also used as substances applied to crops either before or after harvest to protect the commodity from deterioration during storage and transport.

Pesticides can be classified by target organism (e.g., herbicides, insecticides, fungicides, rodenticides, and pediculicides - see table), chemical structure (e.g., organic, inorganic, synthetic, or biological (biopesticide), although the distinction can sometimes blur), and physical state (e.g. gaseous (fumigant)). Biopesticides include microbial pesticides and biochemical pesticides. Plant-derived pesticides, or "botanicals", have been developing quickly. These include the pyrethroids, rotenoids, nicotinoids, and a fourth group that includes strychnine and scilliroside.

Many pesticides can be grouped into chemical families. Prominent insecticide families include organochlorines, organophosphates, and carbamates. Organochlorine hydrocarbons (e.g., DDT) could be separated into dichlorodiphenylethanes, cyclodiene compounds, and other related compounds. They operate by disrupting the sodium/potassium balance of the nerve fiber, forcing the nerve to transmit continuously. Their toxicities vary greatly, but they have been phased out because of their persistence and potential to bioaccumulate. Organophosphate and carbamates largely replaced organochlorines. Both operate through inhibiting the enzyme acetylcholinesterase, allowing acetylcholine to transfer nerve impulses indefinitely and causing a variety of symptoms such as weakness or paralysis. Organophosphates are quite toxic to vertebrates, and have in some cases been replaced by less toxic carbamates. Thiocarbamate and dithiocarbamates are subclasses of carbamates. Prominent families of herbicides include phenoxy and benzoic acid herbicides (e.g. 2,4-D), triazines (e.g., atrazine), ureas (e.g., diuron), and Chloroacetanilides (e.g., alachlor). Phenoxy compounds tend to selectively kill broad-leaf weeds rather than grasses. The phenoxy and benzoic acid herbicides function similar to plant growth hormones, and grow cells without normal cell division, crushing the plant's nutrient transport system. Triazines interfere with photosynthesis. Many commonly used pesticides are not included in these families, including glyphosate.

Pesticides can be classified based upon their biological mechanism function or application method. Most pesticides work by poisoning pests. A systemic pesticide moves inside a plant following absorption by the plant. With insecticides and most fungicides, this movement is usually upward (through the xylem) and outward. Increased efficiency may be a result. Systemic insecticides, which poison pollen and nectar in the flowers, may kill bees and other needed pollinators.

In 2009, the development of a new class of fungicides called paldoxins was announced. These work by taking advantage of natural defense chemicals released by plants called phytoalexins, which fungi then detoxify using enzymes. The paldoxins inhibit the fungi's detoxification enzymes. They are believed to be safer and greener.

Uses

Pesticides are used to control organisms that are considered to be harmful. For example, they are used to kill mosquitoes that can transmit potentially deadly diseases like West Nile virus, yellow fever, and malaria. They can also kill bees, wasps or ants that can cause allergic reactions. Insecticides can protect animals from illnesses that can be caused by parasites such as fleas. Pesticides can prevent sickness in humans that could be caused by moldy food or diseased produce. Herbicides can be used to clear roadside weeds, trees and brush. They can also kill invasive weeds that may cause environmental damage. Herbicides are commonly applied in ponds and lakes to control algae and plants such as water grasses that can interfere with activities like swimming and fishing and cause the water to look or smell unpleasant. Uncontrolled pests such as termites and mold can damage structures such as houses. Pesticides are used in grocery stores and food storage facilities to manage rodents and insects that infest food such as grain. Each use of a pesticide carries some associated risk. Proper pesticide use decreases these associated risks to a level deemed acceptable by pesticide regulatory agencies such as the United States Environmental Protection Agency (EPA) and the Pest Management Regulatory Agency (PMRA) of Canada.

DDT, sprayed on the walls of houses, is an organochlorine that has been used to fight malaria since the 1950s. Recent policy statements by the World Health Organization have given stronger support to this approach. However, DDT and other organochlorine pesticides have been banned in most countries worldwide because of their persistence in the environment and human toxicity. DDT use is not always effective, as resistance to DDT was identified in Africa as early as 1955, and by 1972 nineteen species of mosquito worldwide were resistant to DDT.

Amount used

In 2006 and 2007, the world used approximately 2.4 megatonnes (5.3×10^9 lb) of pesticides, with herbicides constituting the biggest part of the world pesticide use at 40%, followed by insecticides (17%) and fungicides (10%). In 2006 and 2007 the U.S. used approximately 0.5 megatonnes (1.1×10^9 lb) of pesticides, accounting for 22% of the world total, including 857 million pounds (389 kt) of conventional pesticides, which are used in the agricultural sector (80% of conventional pesticide use) as well as the industrial, commercial, governmental and home & garden sectors. Pesticides are also found in majority of U.S. households with 78 million out of the 105.5 million households indicating that they use some form of pesticide. As of 2007, there were more than 1,055 active ingredients registered as pesticides, which yield over 20,000 pesticide products that are marketed in the United States.

The US used some 1 kg (2.2 pounds) per hectare of arable land compared with: 4.7 kg in China, 1.3 kg in the UK, 0.1 kg in Cameroon, 5.9 kg in Japan and 2.5 kg in Italy. Insecticide use in the US has declined by more than half since 1980, (.6%/yr) mostly due to the near phase-out of organo-phosphates. In corn fields, the decline was even steeper, due to the switchover to transgenic Bt corn.

For the global market of crop protection products, market analysts forecast revenues of over 52 billion US$ in 2019.

Benefits

Pesticides can save farmers' money by preventing crop losses to insects and other pests; in the U.S., farmers get an estimated fourfold return on money they spend on pesticides. One study found that not using pesticides reduced crop yields by about 10%. Another study, conducted in 1999, found that a ban on pesticides in the United States may result in a rise of food prices, loss of jobs, and an increase in world hunger.

There are two levels of benefits for pesticide use, primary and secondary. Primary benefits are direct gains from the use of pesticides and secondary benefits are effects that are more long-term.

Primary benefits

1. Controlling pests and plant disease vectors

    o   Improved crop/livestock yields

    o   Improved crop/livestock quality

    o   Invasive species controlled

2. Controlling human/livestock disease vectors and nuisance organisms

    o   Human lives saved and suffering reduced

    o   Animal lives saved and suffering reduced

    o   Diseases contained geographically

3. Controlling organisms that harm other human activities and structures

    o   Drivers view unobstructed

    o   Tree/brush/leaf hazards prevented

    o   Wooden structures protected

Monetary

Every dollar ($1) that is spent on pesticides for crops yields four dollars ($4) in crops saved. This means based that, on the amount of money spent per year on pesticides, $10 billion, there is an additional $40 billion savings in crop that would be lost due to damage by insects and weeds. In general, farmers benefit from having an increase in crop yield and from being able to grow a variety of crops throughout the year. Consumers of agricultural products also benefit from being able to afford the vast quantities of produce available year-round. The general public also benefits from the use of pesticides for the control of insect-borne diseases and illnesses, such as malaria. The use of pesticides creates a large job market within the agrichemical sector.

Costs

On the cost side of pesticide use there can be costs to the environment, costs to human health, as well as costs of the development and research of new pesticides.

Health Effects

A sign warning about potential pesticide exposure.

Pesticides may cause acute and delayed health effects in people who are exposed. Pesticide exposure can cause a variety of adverse health effects, ranging from simple irritation of the skin and eyes to more severe effects such as affecting the nervous system, mimicking hormones causing reproductive problems, and also causing cancer. A 2007 systematic review found that "most studies on non-Hodgkin lymphoma and leukemia showed positive associations with pesticide exposure" and thus concluded that cosmetic use of pesticides should be decreased. There is substantial evidence of associations between organophosphate insecticide exposures and neurobehavioral alterations. Limited evidence also exists for other negative outcomes from pesticide exposure including neurological, birth defects, and fetal death.

The American Academy of Pediatrics recommends limiting exposure of children to pesticides and using safer alternatives:

The World Health Organization and the UN Environment Programme estimate that each year, 3 million workers in agriculture in the developing world experience severe poisoning from pesticides, about 18,000 of whom die. Owing to inadequate regulation and safety precautions, 99% of pesticide related deaths occur in developing countries that account for only 25% of pesticide usage. According to one study, as many as 25 million workers in developing countries may suffer mild pesticide poisoning yearly. There are several careers aside from agriculture that may also put individuals at risk of health effects from pesticide exposure including pet groomers, groundskeepers, and fumigators.

One study found pesticide self-poisoning the method of choice in one third of suicides worldwide, and recommended, among other things, more restrictions on the types of pesticides that are most harmful to humans.

A 2014 epidemiological review found associations between autism and exposure to certain pesticides, but noted that the available evidence was insufficient to conclude that the relationship was causal.

Environmental Effect

Pesticide use raises a number of environmental concerns. Over 98% of sprayed insecticides and 95% of herbicides reach a destination other than their target species, including non-target species, air, water and soil. Pesticide drift occurs when pesticides suspended in the air as particles are carried by wind to other areas, potentially contaminating them. Pesticides are one of the causes of water pollution, and some pesticides are persistent organic pollutants and contribute to soil contamination.

In addition, pesticide use reduces biodiversity, contributes to pollinator decline, destroys habitat (especially for birds), and threatens endangered species. Pests can develop a resistance to the pesticide (pesticide resistance), necessitating a new pesticide. Alternatively a greater dose of the pesticide can be used to counteract the resistance, although this will cause a worsening of the ambient pollution problem.

Since chlorinated hydrocarbon pesticides dissolve in fats and are not excreted, organisms tend to retain them almost indefinitely. Biological magnification is the process whereby these chlorinated hydrocarbons (pesticides) are more concentrated at each level of the food chain. Among marine animals, pesticide concentrations are higher in carnivorous fishes, and even more so in the fish-eating birds and mammals at the top of the ecological pyramid. Global distillation is the process whereby pesticides are transported from warmer to colder regions of the Earth, in particular the Poles and mountain tops. Pesticides that evaporate into the atmosphere at relatively high temperature can be carried considerable distances (thousands of kilometers) by the wind to an area of lower temperature, where they condense and are carried back to the ground in rain or snow.

In order to reduce negative impacts, it is desirable that pesticides be degradable or at least quickly deactivated in the environment. Such loss of activity or toxicity of pesticides is due to both innate chemical properties of the compounds and environmental processes or conditions. For example, the presence of halogens within a chemical structure often slows down degradation in an aerobic environment. Adsorption to soil may retard pesticide movement, but also may reduce bioavailability to microbial degraders.

Economics

Harm	Annual US cost
Public health	$1.1 billion
Pesticide resistance in pest	$1.5 billion
Crop losses caused by pesticides	$1.4 billion
Bird losses due to pesticides	$2.2 billion
Groundwater contamination	$2.0 billion
Other costs	$1.4 billion
Total costs	**$9.6 billion**

Human health and environmental cost from pesticides in the United States is estimated at $9.6 billion offset by about $40 billion in increased agricultural production:

Additional costs include the registration process and the cost of purchasing pesticides. The registration process can take several years to complete (there are 70 different types of field test) and can cost $50–70 million for a single pesticide. Annually the United States spends $10 billion on pesticides.

Alternatives

Alternatives to pesticides are available and include methods of cultivation, use of biological pest controls (such as pheromones and microbial pesticides), genetic engineering, and methods of interfering with insect breeding. Application of composted yard waste has also been used as a way of controlling pests. These methods are becoming increasingly popular and often are safer than traditional chemical pesticides. In addition, EPA is registering reduced-risk conventional pesticides in increasing numbers.

Cultivation practices include polyculture (growing multiple types of plants), crop rotation, planting crops in areas where the pests that damage them do not live, timing planting according to when pests will be least problematic, and use of trap crops that attract pests away from the real crop. In the U.S., farmers have had success controlling insects by spraying with hot water at a cost that is about the same as pesticide spraying.

Release of other organisms that fight the pest is another example of an alternative to pesticide use. These organisms can include natural predators or parasites of the pests. Biological pesticides based on entomopathogenic fungi, bacteria and viruses cause disease in the pest species can also be used.

Interfering with insects' reproduction can be accomplished by sterilizing males of the target species and releasing them, so that they mate with females but do not produce offspring. This technique was first used on the screwworm fly in 1958 and has since been used with the medfly, the tsetse fly, and the gypsy moth. However, this can be a costly, time consuming approach that only works on some types of insects.

Agroecology emphasize nutrient recycling, use of locally available and renewable resources, adaptation to local conditions, utilization of microenvironments, reliance on indigenous knowledge and yield maximization while maintaining soil productivity. Agroecology also emphasizes empowering people and local communities to contribute to development, and encouraging "multi-directional" communications rather than the conventional "top-down" method.

Push Pull Strategy

The term "push-pull" was established in 1987 as an approach for integrated pest management (IPM). This strategy uses a mixture of behavior-modifying stimuli to manipulate the distribution and abundance of insects. "Push" means the insects are repelled or deterred away from whatever resource that is being protected. "Pull" means that certain stimuli (semiochemical stimuli, pheromones, food additives, visual stimuli, genetically altered plants, etc.) are used to attract pests to trap crops where they will be killed. There are numerous different components involved in order to implement a Push-Pull Strategy in IPM.

Many case studies testing the effectiveness of the push-pull approach have been done across the world. The most successful push-pull strategy was developed in Africa for subsistence farming. Another successful case study was performed on the control of *Helicoverpa* in cotton crops in Australia. In Europe, the Middle East, and the United States, push-pull strategies were successfully used in the controlling of *Sitona lineatus* in bean fields.

Some advantages of using the push-pull method are less use of chemical or biological materials and better protection against insect habituation to this control method. Some disadvantages of the push-pull strategy is that if there is a lack of appropriate knowledge of behavioral and chemical ecology of the host-pest interactions then this method becomes unreliable. Furthermore, because the push-pull method is not a very popular method of IPM operational and registration costs are higher.

Effectiveness

Some evidence shows that alternatives to pesticides can be equally effective as the use of chemicals. For example, Sweden has halved its use of pesticides with hardly any reduction in crops.In Indonesia, farmers have reduced pesticide use on rice fields by 65% and experienced a 15% crop increase. A study of Maize fields in northern Florida found that the application of composted yard waste with high carbon to nitrogen ratio to agricultural fields was highly effective at reducing the population of plant-parasitic nematodes and increasing crop yield, with yield increases ranging from 10% to 212%; the observed effects were long-term, often not appearing until the third season of the study.

However, pesticide resistance is increasing. In the 1940s, U.S. farmers lost only 7% of their crops to pests. Since the 1980s, loss has increased to 13%, even though more pesticides are being used. Between 500 and 1,000 insect and weed species have developed pesticide resistance since 1945.

Types

Pesticides are often referred to according to the type of pest they control. Pesticides can also be considered as either biodegradable pesticides, which will be broken down by microbes and other living beings into harmless compounds, or persistent pesticides, which may take months or years before they are broken down: it was the persistence of DDT, for example, which led to its accumulation in the food chain and its killing of birds of prey at the top of the food chain. Another way to think about pesticides is to consider those that are chemical pesticides or are derived from a common source or production method.

Some examples of chemically-related pesticides are:

Organophosphate Pesticides

Organophosphates affect the nervous system by disrupting, acetylcholinesterase activity, the enzyme that regulates acetylcholine, a neurotransmitter. Most organophosphates are insecticides. They were developed during the early 19th century, but their effects on insects, which are similar to their effects on humans, were discovered in 1932. Some are very poisonous. However, they usually are not persistent in the environment.

Carbamate Pesticides

Carbamate pesticides affect the nervous system by disrupting an enzyme that regulates acetylcholine, a neurotransmitter. The enzyme effects are usually reversible. There are several subgroups within the carbamates.

Organochlorine Insecticides

They were commonly used in the past, but many have been removed from the market due to their health and environmental effects and their persistence (e.g., DDT, chlordane, and toxaphene).

Pyrethroid Pesticides

They were developed as a synthetic version of the naturally occurring pesticide pyrethrin, which is found in chrysanthemums. They have been modified to increase their stability in the environment. Some synthetic pyrethroids are toxic to the nervous system.

Sulfonylurea Herbicides

The following sulfonylureas have been commercialized for weed control: amidosulfuron, azimsulfuron, bensulfuron-methyl, chlorimuron-ethyl, ethoxysulfuron, flazasulfuron, flupyrsulfuron-methyl-sodium, halosulfuron-methyl, imazosulfuron, nicosulfuron, oxasulfuron, primisulfuron-methyl, pyrazosulfuron-ethyl, rimsulfuron, sulfometuron-methyl Sulfosulfuron, terbacil, bispyribac-sodium, cyclosulfamuron, and pyrithiobac-sodium. Nicosulfuron, triflusulfuron methyl, and chlorsulfuron are broad-spectrum herbicides that kill plants by inhibiting the enzyme acetolactate synthase. In the 1960s, more than 1 kg/ha (0.89 lb/acre) crop protection chemical was typically applied, while sulfonylureates allow as little as 1% as much material to achieve the same effect.

Biopesticides

Biopesticides are certain types of pesticides derived from such natural materials as animals, plants, bacteria, and certain minerals. For example, canola oil and baking soda have pesticidal applications and are considered biopesticides. Biopesticides fall into three major classes:

- Microbial pesticides which consist of bacteria, entomopathogenic fungi or viruses (and sometimes includes the metabolites that bacteria or fungi produce). Entomopathogenic nematodes are also often classed as microbial pesticides, even though they are multi-cellular.

- Biochemical pesticides or herbal pesticides are naturally occurring substances that control (or monitor in the case of pheromones) pests and microbial diseases.

- Plant-incorporated protectants (PIPs) have genetic material from other species incorporated into their genetic material (*i.e.* GM crops). Their use is controversial, especially in many European countries.

Classified by Type of Pest

Pesticides that are related to the type of pests are:

Type	Action
Algicides	Control algae in lakes, canals, swimming pools, water tanks, and other sites
Antifouling agents	Kill or repel organisms that attach to underwater surfaces, such as boat bottoms
Antimicrobials	Kill microorganisms (such as bacteria and viruses)
Attractants	Attract pests (for example, to lure an insect or rodent to a trap). (However, food is not considered a pesticide when used as an attractant.)
Biopesticides	Biopesticides are certain types of pesticides derived from such natural materials as animals, plants, bacteria, and certain minerals
Biocides	Kill microorganisms
Disinfectants and sanitizers	Kill or inactivate disease-producing microorganisms on inanimate objects
Fungicides	Kill fungi (including blights, mildews, molds, and rusts)
Fumigants	Produce gas or vapor intended to destroy pests in buildings or soil
Herbicides	Kill weeds and other plants that grow where they are not wanted
Insecticides	Kill insects and other arthropods
Miticides	Kill mites that feed on plants and animals
Microbial pesticides	Microorganisms that kill, inhibit, or out compete pests, including insects or other microorganisms
Molluscicides	Kill snails and slugs
Nematicides	Kill nematodes (microscopic, worm-like organisms that feed on plant roots)
Ovicides	Kill eggs of insects and mites
Pheromones	Biochemicals used to disrupt the mating behavior of insects
Repellents	Repel pests, including insects (such as mosquitoes) and birds
Rodenticides	Control mice and other rodents

Further Types of Pesticides

The term pesticide also include these substances:

Defoliants : Cause leaves or other foliage to drop from a plant, usually to facilitate harvest. Desiccants : Promote drying of living tissues, such as unwanted plant tops. Insect growth regulators : Disrupt the molting, maturity from pupal stage to adult, or other life processes of insects. Plant growth regulators : Substances (excluding fertilizers or other plant nutrients) that alter the expected growth, flowering, or reproduction rate of plants.

Regulation

International

In most countries, pesticides must be approved for sale and use by a government agency.

In Europe, recent EU legislation has been approved banning the use of highly toxic pesticides including those that are carcinogenic, mutagenic or toxic to reproduction, those that are endocrine-disrupting, and those that are persistent, bioaccumulative and toxic (PBT) or very persistent and very bioaccumulative (vPvB). Measures were approved to improve the general safety of pesticides across all EU member states.

Though pesticide regulations differ from country to country, pesticides, and products on which they were used are traded across international borders. To deal with inconsistencies in regulations among countries, delegates to a conference of the United Nations Food and Agriculture Organization adopted an International Code of Conduct on the Distribution and Use of Pesticides in 1985 to create voluntary standards of pesticide regulation for different countries. The Code was updated in 1998 and 2002. The FAO claims that the code has raised awareness about pesticide hazards and decreased the number of countries without restrictions on pesticide use.

Three other efforts to improve regulation of international pesticide trade are the United Nations London Guidelines for the Exchange of Information on Chemicals in International Trade and the United Nations Codex Alimentarius Commission. The former seeks to implement procedures for ensuring that prior informed consent exists between countries buying and selling pesticides, while the latter seeks to create uniform standards for maximum levels of pesticide residues among participating countries. Both initiatives operate on a voluntary basis.

Pesticides safety education and pesticide applicator regulation are designed to protect the public from pesticide misuse, but do not eliminate all misuse. Reducing the use of pesticides and choosing less toxic pesticides may reduce risks placed on society and the environment from pesticide use. Integrated pest management, the use of multiple approaches to control pests, is becoming widespread and has been used with success in countries such as Indonesia, China, Bangladesh, the U.S., Australia, and Mexico. IPM attempts to recognize the more widespread impacts of an action on an ecosystem, so that natural balances are not upset. New pesticides are being developed, including biological and botanical derivatives and alternatives that are thought to reduce health and environmental risks. In addition, applicators are being encouraged to consider alternative controls and adopt methods that reduce the use of chemical pesticides.

Pesticides can be created that are targeted to a specific pest's lifecycle, which can be environmentally more friendly. For example, potato cyst nematodes emerge from their protective cysts in response to a chemical excreted by potatoes; they feed on the potatoes and damage the crop. A similar chemical can be applied to fields early, before the potatoes are planted, causing the nematodes to emerge early and starve in the absence of potatoes.

United States

Preparation for an application of hazardous herbicide in USA.

In the United States, the Environmental Protection Agency (EPA) is responsible for regulating pesticides under the Federal Insecticide, Fungicide, and Rodenticide Act (FIFRA) and the Food Quality Protection Act (FQPA). Studies must be conducted to establish the conditions in which the material is safe to use and the effectiveness against the intended pest(s). The EPA regulates pesticides to ensure that these products do not pose adverse effects to humans or the environment. Pesticides produced before November 1984 continue to be reassessed in order to meet the current scientific and regulatory standards. All registered pesticides are reviewed every 15 years to ensure they meet the proper standards. During the registration process, a label is created. The label contains directions for proper use of the material in addition to safety restrictions. Based on acute toxicity, pesticides are assigned to a Toxicity Class.

Some pesticides are considered too hazardous for sale to the general public and are designated restricted use pesticides. Only certified applicators, who have passed an exam, may purchase or supervise the application of restricted use pesticides. Records of sales and use are required to be maintained and may be audited by government agencies charged with the enforcement of pesticide regulations. These records must be made available to employees and state or territorial environmental regulatory agencies.

The EPA regulates pesticides under two main acts, both of which amended by the Food Quality Protection Act of 1996. In addition to the EPA, the United States Department of Agriculture (USDA) and the United States Food and Drug Administration (FDA) set standards for the level of pesticide residue that is allowed on or in crops. The EPA looks at what the potential human health and environmental effects might be associated with the use of the pesticide.

In addition, the U.S. EPA uses the National Research Council's four-step process for human health risk assessment: (1) Hazard Identification, (2) Dose-Response Assessment, (3) Exposure Assessment, and (4) Risk Characterization.

Recently Kaua'i County (Hawai'i) passed Bill No. 2491 to add an article to Chapter 22 of the county's code relating to pesticides and GMOs. The bill strengthens protections of local communities in Kaua'i where many large pesticide companies test their products.

History

Since before 2000 BC, humans have utilized pesticides to protect their crops. The first known pesticide was elemental sulfur dusting used in ancient Sumer about 4,500 years ago in ancient Mesopotamia. The Rig Veda, which is about 4,000 years old, mentions the use of poisonous plants for pest control. By the 15th century, toxic chemicals such as arsenic, mercury, and lead were being applied to crops to kill pests. In the 17th century, nicotine sulfate was extracted from tobacco leaves for use as an insecticide. The 19th century saw the introduction of two more natural pesticides, pyrethrum, which is derived from chrysanthemums, and rotenone, which is derived from the roots of tropical vegetables. Until the 1950s, arsenic-based pesticides were dominant. Paul Müller discovered that DDT was a very effective insecticide. Organochlorines such as DDT were dominant, but they were replaced in the U.S. by organophosphates and carbamates by 1975. Since then, pyrethrin compounds have become the dominant insecticide. Herbicides became common in the 1960s, led by "triazine and other nitrogen-based compounds, carboxylic acids such as 2,4-dichlorophenoxyacetic acid, and glyphosate".

The first legislation providing federal authority for regulating pesticides was enacted in 1910; however, decades later during the 1940s manufacturers began to produce large amounts of synthetic pesticides and their use became widespread. Some sources consider the 1940s and 1950s to have been the start of the "pesticide era." Although the U.S. Environmental Protection Agency was established in 1970 and amendments to the pesticide law in 1972, pesticide use has increased 50-fold since 1950 and 2.3 million tonnes (2.5 million short tons) of industrial pesticides are now used each year. Seventy-five percent of all pesticides in the world are used in developed countries, but use in developing countries is increasing. A study of USA pesticide use trends through 1997 was published in 2003 by the National Science Foundation's Center for Integrated Pest Management.

In the 1960s, it was discovered that DDT was preventing many fish-eating birds from reproducing, which was a serious threat to biodiversity. Rachel Carson wrote the best-selling book *Silent Spring* about biological magnification. The agricultural use of DDT is now banned under the Stockholm Convention on Persistent Organic Pollutants, but it is still used in some developing nations to prevent malaria and other tropical diseases by spraying on interior walls to kill or repel mosquitoes.

Biological Pest Control

Biological control is a method of controlling pests such as insects, mites, weeds and plant diseases using other organisms. It relies on predation, parasitism, herbivory, or other natural mechanisms, but typically also involves an active human management role. It can be an important component of integrated pest management (IPM) programs.

There are three basic types of biological pest control strategies: importation (sometimes called classical biological control), in which a natural enemy of a pest is introduced in the hope of achieving control; augmentation, in which locally-occurring natural enemies are bred and released to improve control; and conservation, in which measures are taken to increase natural enemies, such as by planting nectar-producing crop plants in the borders of rice fields.

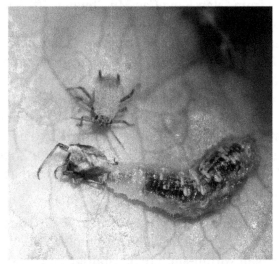

Syrphus hoverfly larva feeding on aphids

Parasitic wasp *Cotesia congregata* on tobacco hornworm *Manduca sexta*

Natural enemies of insect pests, also known as biological control agents, include predators, parasitoids, and pathogens. Biological control agents of plant diseases are most often referred to as antagonists. Biological control agents of weeds include seed predators, herbivores and plant pathogens.

Biological control can have side-effects on biodiversity through predation, parasitism, pathogenicity, competition, or other attacks on non-target species, especially when a species is introduced without thorough understanding of the possible consequences.

History of Biological Control

The term "biological control" was first used by Harry Scott Smith at the 1919 meeting of the Pacific Slope Branch of the American Association of Economic Entomologists, at the Mission Inn in downtown Riverside, California; and later defined by P. DeBach and K. S. Hagen in 1964. However, the practice has previously been used for centuries. The first report of the use of an insect species to control an insect pest comes from "Nan Fang Cao Mu Zhuang" (南方草木 *Plants of the Southern Regions*) (ca. 304 AD), which is attributed to Western Jin dynasty botanist *Ji Han* (含, 263-307), in which it is mentioned that "*Jiaozhi people sell ants and their nests attached to twigs looking like thin cotton envelopes, the reddish-yellow ant being larger than normal. Without such ants, southern citrus fruits will be severely insect-damaged*". The ants used are known as *huang gan* (*huang* = yellow, *gan* = citrus) ants (*Oecophylla smaragdina*). This practice has later also been reported by *Ling Biao Lu Yi* (late Tang Dynasty or Early Five Dynasties), in "Ji Le Pian" by *Zhuang Jisu* (Southern Song Dynasty), in the "Book of Tree Planting" by *Yu Zhen Mu* (Ming Dynasty), in the book "Guangdong Xing Yu" (17th century), "Lingnan" by *Wu Zhen Fang* (Qing Dynasty), in "Nanyue Miscellanies" by *Li Diao Yuan*, and others, which shows that this practice has obviously perdured for a very long time.

The use of Biological control techniques as we know them today started to emerge in the 1870s. During this decade, in the USA, The Missouri State Entomologist C. V. Riley and the Illinois State Entomologist W. LeBaron began within-state redistribution of parasitoids to control crop pests. The first international shipment of an insect as biological control agent was made by Charles V. Riley in 1873, shipping to France the predatory mites *Tyroglyphus phylloxera* to help fight the grapevine phylloxera (*Daktulosphaira vitifoliae*) that was destroying grapevines in France. The United States Department of Agriculture (USDA) initiated research in classical biological control following the establishment of the Division of Entomology in 1881, with C. V. Riley as Chief. The first importation of a parasitoid into the United States was this of *Cotesia glomerata* in 1883-1884, imported from Europe to control the imported cabbage white butterfly, *Pieris rapae*. In 1888-1889 the vedalia beetle, *Rodolia cardinalis*, which is a ladybug, was imported from Australia and introduced into California to control the cottony cushion scale, *Icerya purchasi*, which had become a major problem for the newly developed citrus industry in California, and by the end of 1889 the cottony cushion scale population had already declined. This great success led to further introductions of beneficial insects into the USA.

In 1905 the USDA initiated its first large-scale biological control program, sending entomologists to Europe and Japan to look for natural enemies of the gypsy moth, *Lymantria dispar dispar*, and brown-tail moth, *Euproctis chrysorrhoea*, invasive pests of trees and shrubs. As a result, nine species of parasitoid of gypsy moth, seven of brown-tail moth, and two predators for both moths became established in the USA. Although the gypsy moth was not fully controlled by these natural enemies, the frequency, duration, and severity of its outbreaks were reduced and the program was regarded as successful. This program also led to the development of many concepts, principles, and procedures for the implementation of biological control programs.

The first reported case of a classical biological control attempt in Canada involves the hymenopteran parasitoid *Trichogramma minutum*. Individuals were caught in New York State and released in Ontario gardens in 1882 by William Saunders, trained chemist and first Director of the Dominion Experimental Farms, for controlling the imported currantworm *Nematus ribesii*. Between 1884 and 1908, the first Dominion Entomologist, James Fletcher, continued introductions of other parasitoids and pathogens for the control of pests in Canada.

Types of Biological Pest Control

There are three basic biological pest control strategies: importation (classical biological control), augmentation and conservation.

Importation

Rodolia cardinalis, the vedalia beetle, was imported to Australia in the 19th century, successfully controlling cottony cushion scale.

Importation or classical biological control involves the introduction of a pest's natural enemies to a new locale where they do not occur naturally. Early instances were often unofficial and not based on research, and some introduced species became serious pests themselves.

To be most effective at controlling a pest, a biological control agent requires a colonizing ability which allows it to keep pace with the spatial and temporal disruption of the habitat. Control is greatest if the agent has temporal persistence, so that it can maintain its population even in the temporary absence of the target species, and if it is an opportunistic forager, enabling it to rapidly exploit a pest population.

Joseph Needham noted a Chinese text dating from 304 AD, *Records of the Plants and Trees of the Southern Regions*, by Hsi Han, which describes mandarin oranges protected by large reddish-yellow citrus ants which attack and kill insect pests of the orange trees. The citrus ant (*Oecophylla smaragdina*) was rediscovered in the 20th century, and since 1958 has been used in China to protect orange groves.

One of the earliest successes in the west was in controlling *Icerya purchasi* (cottony cushion scale) in Australia, using a predatory insect *Rodolia cardinalis* (the vedalia beetle). This success was repeated in California using the beetle and a parasitoid fly, *Cryptochaetum iceryae*.

Prickly pear cacti were introduced into Queensland, Australia as ornamental plants. They quickly spread to cover over 25 million hectares of Australia. Two control agents were used to help control the spread of the plant, the cactus moth *Cactoblastis cactorum*, and *Dactylopius* scale insects.

Cactoblastis cactorum larvae feeding on *Opuntia* cacti

Damage from *Hypera postica*, the alfalfa weevil, a serious introduced pest of forage, was substantially reduced by the introduction of natural enemies. 20 years after their introduction the population of weevils in the alfalfa area treated for alfalfa weevil in the Northeastern United States remained 75 percent down.

The invasive species *Alternanthera philoxeroides* (alligator weed) was controlled
in Florida (U.S.) by introducing alligator weed flea beetle.

Alligator weed was introduced to the United States from South America. It takes root in shallow water, interfering with navigation, irrigation, and flood control. The alligator weed flea beetle and two other biological controls were released in Florida, enabling the state to ban the use of herbicides to control alligator weed three years later. Another aquatic weed, the giant salvinia (*Salvinia molesta*) is a serious pest, covering waterways, reducing water flow and harming native species. Control with the salvinia weevil (*Cyrtobagous salviniae*) is effective in warm climates, and in Zimbabwe, a 99% control of the weed was obtained over a two-year period.

Small commercially reared parasitoidal wasps, *Trichogramma ostriniae*, provide limited and erratic control of the European corn borer (*Ostrinia nubilalis*), a serious pest. Careful formulations of the bacterium *Bacillus thuringiensis* are more effective.

The population of *Levuana iridescens*, the Levuana moth, a serious coconut pest in Fiji, was brought under control by a classical biological control program in the 1920s.

Augmentation

Hippodamia convergens, the convergent lady beetle, is commonly sold for biological control of aphids.

Augmentation involves the supplemental release of natural enemies, boosting the naturally occurring population. In inoculative release, small numbers of the control agents are released at intervals to allow them to reproduce, in the hope of setting up longer-term control, and thus keeping the pest down to a low level, constituting prevention rather than cure. In inundative release, in contrast, large numbers are released in the hope of rapidly reducing a damaging pest population, correcting a problem that has already arisen. Augmentation can be effective, but is not guaranteed to work, and relies on understanding of the situation.

An example of inoculative release occurs in greenhouse production of several crops. Periodic releases of the parasitoid, *Encarsia formosa*, are used to control greenhouse whitefly, while the predatory mite *Phytoseiulus persimilis* is used for control of the two-spotted spider mite.

The egg parasite *Trichogramma* is frequently released inundatively to control harmful moths. Similarly, *Bacillus thuringiensis* and other microbial insecticides are similarly used in large enough quantities for a rapid effect. Recommended release rates for *Trichogramma* in vegetable or field crops range from 5,000 to 200,000 per acre (1 to 50 per square metre) per week according to the level of pest infestation. Similarly, entomopathogenic nematodes are released at rates of millions and even billions per acre for control of certain soil-dwelling insect pests.

Conservation

The conservation of existing natural enemies in an environment is the third method of biological pest control. Natural enemies are already adapted to the habitat and to the target pest, and their conservation can be simple and cost-effective, as when nectar-producing crop plants are grown in the borders of rice fields. These provide nectar to support parasitoids and predators of planthopper pests and have been demonstrated to be so effective (reducing pest densities by 10- or even

100-fold) that farmers sprayed 70% less insecticides, enjoyed yields boosted by 5%, and this led to an economic advantage of 7.5%. Predators of aphids were similarly found to be present in tussock grasses by field boundary hedges in England, but they spread too slowly to reach the centres of fields. Control was improved by planting a metre-wide strip of tussock grasses in field centres, enabling aphid predators to overwinter there.

An inverted flowerpot filled with straw to attract earwigs

Cropping systems can be modified to favor natural enemies, a practice sometimes referred to as habitat manipulation. Providing a suitable habitat, such as a shelterbelt, hedgerow, or beetle bank where beneficial insects can live and reproduce, can help ensure the survival of populations of natural enemies. Things as simple as leaving a layer of fallen leaves or mulch in place provides a suitable food source for worms and provides a shelter for insects, in turn being a food source for such beneficial mammals as hedgehogs and shrews. Compost piles and stacks of wood can provide shelter for invertebrates and small mammals. Long grass and ponds support amphibians. Not removing dead annuals and non-hardy plants in the autumn allows insects to make use of their hollow stems during winter. In California, prune trees are sometimes planted in grape vineyards to provide an improved overwintering habitat or refuge for a key grape pest parasitoid. The providing of artificial shelters in the form of wooden caskets, boxes or flowerpots is also sometimes undertaken, particularly in gardens, to make a cropped area more attractive to natural enemies. For example, earwigs are natural predators which can be encouraged in gardens by hanging upside-down flowerpots filled with straw or wood wool. Green lacewings can be encouraged by using plastic bottles with an open bottom and a roll of cardboard inside. Birdhouses enable insectivorous birds to nest; the most useful birds can be attracted by choosing an opening just large enough for the desired species.

Biological Control Agents

Predators

Predators are mainly free-living species that directly consume a large number of prey during their whole lifetime. Ladybugs, and in particular their larvae which are active between May and July in the northern hemisphere, are voracious predators of aphids, and also consume mites, scale insects and small caterpillars. The spotted lady beetle (*Coleomegilla maculata*) is also able to feed on the eggs and larvae of the Colorado potato beetle (*Leptinotarsa decemlineata*).

Lacewings are available from biocontrol dealers.

The larvae of many hoverfly species principally feed upon greenfly (aphids), one larva devouring up to 400 in its lifetime. Their effectiveness in commercial crops has not been studied.

Several species of entomopathogenic nematode are important predators of insect and other invertebrate pests. *Phasmarhabditis hermaphrodita* is a microscopic nematode that kills slugs. Its complex life cycle include a free-living, infective stage in the soil where it becomes associated with a pathogenic bacteria such as *Moraxella osloensis*. The nematode enters the slug through the posterior mantle region, thereafter feeding and reproducing inside, but it is the bacteria that kill the slug. The nematode is available commercially in Europe and is applied by watering onto moist soil.

Predatory *Polistes* wasp looking for bollworms or other caterpillars on a cotton plant

Species used to control spider mites include the predatory mites *Phytoseiulus persimilis*, *Neoseilus californicus*, and *Amblyseius cucumeris*, the predatory midge *Feltiella acarisuga*, and a ladybird *Stethorus punctillum*. The bug *Orius insidiosus* has been successfully used against the two-spotted spider mite and the western flower thrips (*Frankliniella occidentalis*).

Parasitoids

Parasitoids lay their eggs on or in the body of an insect host, which is then used as a food for developing larvae. The host is ultimately killed. Most insect parasitoids are wasps or flies, and may have a very narrow host range. The most important groups are the ichneumonid wasps, which prey mainly on caterpillars of butterflies and moths; braconid wasps, which attack caterpillars and a wide range of other insects including greenfly; chalcid wasps, which parasitize eggs and larvae of greenfly, whitefly, cabbage caterpillars, and scale insects; and tachinid flies, which parasitize a wide range of insects including caterpillars, adult and larval beetles, and true bugs.

Encarsia formosa was one of the first biological control agents developed.

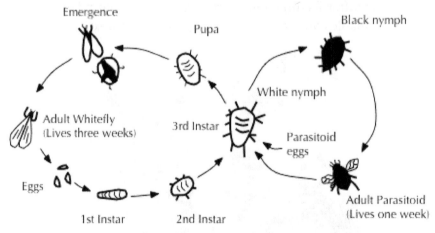

Life cycles of Greenhouse whitefly and its parasitoid wasp *Encarsia formosa*

Encarsia formosa is a small predatory chalcid wasp which is a parasitoid of whitefly, a sap-feeding insect which can cause wilting and black sooty moulds in glasshouse vegetable and ornamental crops. It is most effective when dealing with low level infestations, giving protection over a long period of time. The wasp lays its eggs in young whitefly 'scales', turning them black as the parasite larvae pupates. *Gonatocerus ashmeadi* (Hymenoptera: Mymaridae) has been introduced to control the glassy-winged sharpshooter *Homalodisca vitripennis* (Hemipterae: Cicadellidae) in French Polynesia and has successfully controlled ~95% of the pest density.

Parasitoids are among the most widely used biological control agents. Commercially, there are two types of rearing systems: short-term daily output with high production of parasitoids per day, and long-term low daily output with a range in production of 4-1000million female parasitoids per week. Larger production facilities produce on a yearlong basis, whereas some facilities produce only seasonally. Rearing facilities are usually a significant distance from where the agents are to be used in the field, and transporting the parasitoids from the point of production to the point of use can pose problems. Shipping conditions can be too hot, and even vibrations from planes or trucks can adversely affect parasitoids.

Pathogens

Pathogenic micro-organisms include bacteria, fungi, and viruses. They kill or debilitate their host and are relatively host-specific. Various microbial insect diseases occur naturally, but may also be used as biological pesticides. When naturally occurring, these outbreaks are density-dependent in that they generally only occur as insect populations become denser.

Bacteria

Bacteria used for biological control infect insects via their digestive tracts, so they offer only limited options for controlling insects with sucking mouth parts such as aphids and scale insects. *Bacillus thuringiensis* is the most widely applied species of bacteria used for biological control, with at least four sub-species used against Lepidopteran (moth, butterfly), Coleopteran (beetle) and Dipteran (true fly) insect pests. The bacterium is available in sachets of dried spores which are mixed with water and sprayed onto vulnerable plants such as brassicas and fruit trees. *B. thuringiensis* has also been incorporated into crops, making them resistant to these pests and thus reducing the use of pesticides. The bacterium *Paenibacillus popilliae* causes milky spore disease has been found useful in the control of Japanese beetle, killing the larvae. It is very specific to its host species and is harmless to vertebrates and other invertebrates.

Fungi

Entomopathogenic fungi, which cause disease in insects, include at least 14 species that attack aphids. *Beauveria bassiana* is mass-produced and used to manage a wide variety of insect pests including whiteflies, thrips, aphids and weevils. *Lecanicillium* spp. are deployed against white flies, thrips and aphids. *Metarhizium* spp. are used against pests including beetles, locusts and other grasshoppers, Hemiptera, and spider mites. *Paecilomyces fumosoroseus* is effective against white flies, thrips and aphids; *Purpureocillium lilacinus* is used against root-knot nematodes, and 89 *Trichoderma* species against certain plant pathogens. *Trichoderma viride* has been used against Dutch elm disease, and has shown some effect in suppressing silver leaf, a disease of stone fruits caused by the pathogenic fungus *Chondrostereum purpureum*.

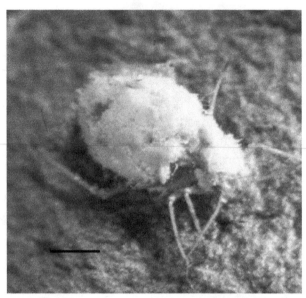

Green peach aphid, a pest in its own right and a vector of plant viruses, killed by the fungus *Pandora neoaphidis* (Zygomycota: Entomophthorales) Scale bar = 0.3 mm.

The fungi *Cordyceps* and *Metacordyceps* are deployed against a wide spectrum of arthropods. *Entomophaga* is effective against pests such as the green peach aphid.

Several members of Chytridiomycota and Blastocladiomycota have been explored as agents of biological control. From Chytridiomycota, *Synchytrium solstitiale* is being considered as a control agent of the yellow star thistle (*Centaurea solstitialis*) in the United States.

Viruses

Baculoviruses are specific to individual insect host species and have been shown to be useful in biological pest control. For example, the Lymantria dispar multicapsid nuclear polyhedrosis virus has been used to spray large areas of forest in North America where larvae of the gypsy moth are causing serious defoliation. The moth larvae are killed by the virus they have eaten and die, the disintegrating cadavers leaving virus particles on the foliage to infect other larvae.

A mammalian virus, the rabbit haemorrhagic disease virus has been introduced to Australia and to New Zealand to attempt to control the European rabbit populations there.

Algae

Lagenidium giganteum is a water-borne mould that parasitizes the larval stage of mosquitoes. When applied to water, the motile spores avoid unsuitable host species and search out suitable mosquito larval hosts. This alga has the advantages of a dormant phase, resistant to desiccation, with slow-release characteristics over several years. Unfortunately, it is susceptible to many chemicals used in mosquito abatement programmes.

Plants

The legume vine *Mucuna pruriens* is used in the countries of Benin and Vietnam as a biological control for problematic *Imperata cylindrica* grass. *Mucuna pruriens* is said not to be invasive outside its cultivated area. *Desmodium uncinatum* can be used in push-pull farming to stop the parasitic plant, *Striga*.

Other Methods

Combined use of Parasitoids and Pathogens

In cases of massive and severe infection of invasive pests, techniques of pest control are often used in combination. An example is the emerald ash borer, *Agrilus planipennis*, an invasive beetle from China, which has destroyed tens of millions of ash trees in its introduced range in North America. As part of the campaign against it, from 2003 American scientists and the Chinese Academy of Forestry searched for its natural enemies in the wild, leading to the discovery of several parasitoid wasps, namely *Tetrastichus planipennisi*, a gregarious larval endoparasitoid,*Oobius agrili*, a solitary, parthenogenic egg parasitoid, and *Spathius agrili*, a gregarious larval ectoparasitoid. These have been introduced and released into the United States of America as a possible biological control of the emerald ash borer. Initial results have shown promise with *Tetrastichus planipennisi* and it is now being released along with *Beauveria bassiana*, a fungal pathogen with known insecticidal properties.

Indirect Control

Pests may be controlled by biological control agents that do not prey directly upon them. For example, the Australian bush fly, *Musca vetustissima*, is a major nuisance pest in Australia, but native decomposers found in Australia are not adapted to feeding on cow dung, which is where bush flies breed. Therefore, the Australian Dung Beetle Project (1965–1985), led by Dr. George Bornemissza of the Commonwealth Scientific and Industrial Research Organisation, released forty-nine species of dung beetle, with the aim of reducing the amount of dung and therefore also the potential breeding sites of the fly.

Side-effects

Biological control can affect biodiversity through predation, parasitism, pathogenicity, competition, or other attacks on non-target species. An introduced control does not always target only the intended pest species; it can also target native species. In Hawaii during the 1940s parasitic wasps were introduced to control a lepidopteran pest and the wasps are still found there today. This may have a negative impact on the native ecosystem, however, host range and impacts need to be studied before declaring their impact on the environment.

Vertebrate animals tend to be generalist feeders, and seldom make good biological control agents; many of the classic cases of "biocontrol gone awry" involve vertebrates. For example, the cane toad (*Bufo marinus*) was intentionally introduced to Australia to control the greyback cane beetle (*Dermolepida albohirtum*), and other pests of sugar cane. 102 toads were obtained from Hawaii and bred in captivity to increase their numbers until they were released into the sugar cane fields of the tropic north in 1935. It was later discovered that the toads could not jump very high and so were unable to eat the cane beetles which stayed up on the upper stalks of the cane plants. However the toad thrived by feeding on other insects and it soon spread very rapidly; it took over native amphibian habitat and brought foreign disease to native toads and frogs, dramatically reducing their populations. Also when it is threatened or handled, the cane toad releases poison from parotoid glands on its shoulders; native Australian species such as goannas, tiger snakes, dingos and northern quolls that attempted to eat the toad were harmed or killed. However, there has been some recent evidence that native predators are adapting, both physiologically and through changing their behaviour, so in the long run, their populations may recover.

Rhinocyllus conicus, a seed-feeding weevil, was introduced to North America to control exotic musk thistle (*Carduus nutans*) and Canadian thistle (*Cirsium arvense*). However the weevil also attacks native thistles, harming such species as the endemic Platte thistle (*Cirsium neomexicanum*) by selecting larger plants (which reduced the gene pool), reducing seed production and ultimately threatening the species' survival.

The small Asian mongoose (*Herpestus javanicus*) was introduced to Hawaii in order to control the rat population. However it was diurnal and the rats emerged at night, and it preyed on the endemic birds of Hawaii, especially their eggs, more often than it ate the rats, and now both rats and mongooses threaten the birds. This introduction was undertaken without understanding the consequences of such an action. No regulations existed at the time, and more careful evaluation should prevent such releases now.

The sturdy and prolific eastern mosquitofish (*Gambusia holbrooki*) is a native of the southeastern United States and was introduced around the world in the 1930s and 40s to feed on mosquito larvae and thus combat malaria. However, it has thrived at the expense of local species, causing a decline of endemic fish and frogs through competition for food resources, as well as through eating their eggs and larvae. In Australia, the mosquitofish is the subject of discussion as to how best to control it; in 1989 it was said that "biological population control is well beyond present capabilities", and this remains the position.

Grower Education

A potential obstacle to the adoption of biological pest control measures is growers sticking to the familiar use of pesticides. It has been claimed that many of the pests that are controlled today using pesticides, actually became pests because pesticide use reduced or eliminated natural predators. A method of increasing grower adoption of biocontrol involves is letting growers learn by doing, for example showing them simple field experiments, having observations of live predation of pests, or collections of parasitised pests. In the Philippines, early season sprays against leaf folder caterpillars were common practice, but growers were asked to follow a 'rule of thumb' of not spraying against leaf folders for the first 30 days after transplanting; participation in this resulted in a reduction of insecticide use by 1/3 and a change in grower perception of insecticide use.

Bird Netting

Bird netting or anti-bird netting is a form of bird pest control. It is a net used to prevent birds from reaching certain areas.

Bird protection netting comes in a variety of shapes and forms, The most common is a small mesh (1 or 2 cm squares) either extruded and bi-oriented polypropylene or woven polyethylene.

Bird netting on grapevines. Most of the side netting has been lifted up for harvesting.

The color most used is black (as the carbon black UV inhibitor offers the best protection against solar rays), but also bird netting may be available in other colors like white (usually white netting is woven or knitted and has an even smaller mesh size as it will serve as a double purpose anti-hail net for the protection of fruits during summer hail storms or late spring during flowering) or green (usually used in home gardening and mostly sold at retail outlets for the DIY farmers).

Professional anti-bird netting comes in jumbo rolls that will offer considerable savings to the farmers or aquaculturalist. Retail chains and local stores will offer smaller packages that fit the backyard gardener´s needs.

Usages

Crop Protection

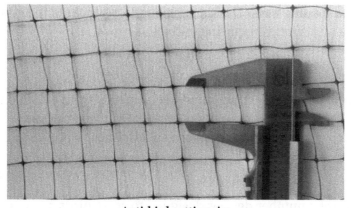

Anti-bird netting size

Bird nets are used to prevent bird damage of vegetable and fruit crops as well as seedlings. Frugivore birds and bats can cause great damages to farmers as they tend to peck one fruit, then go to another, therefore ruining a large percentage of otherwise commercially valuable production.

Once even a small portion is bitten off, that fruit cannot be sold and if harvested (even if there is no bacteria or virus brought by the frugivore) will going into rot or fermentation damaging the rest of the harvested case. Bird protection netting is applied directly on the stand alone trees or espaliers like peaches, pears, apples, grapes, or on the side ventilation windows of growing tunnels as in the case of berries like strawberries, raspberries, blueberries, cranberries.

Fish Protection

Bird netting may be used to protect fisheries and fish wildlife reserves from predator birds. Also in aquaculture (like shrimp and tilapia farms to mention a few), growers need to protect their work and fish crops from marauding birds. These type of birds have usually a larger wing span (seagulls, pelicans, herons, cormorants etc) and a larger mesh size (with individual strands being more resistant as it will be istalled on a cable system crossing the growing ponds). These netting are usually white as to be very visible for the large sea birds will be deterred by the sight of a barrier to their diving into the ponds.

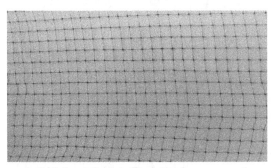

Anti-bird-netting, malla-antipajaros example

Building Protection

Bird netting is one of the most effective and long lasting ways of bird proofing buildings and other structures against all urban bird species. It provides a discreet and impenetrable barrier that protects premises without harming the birds. Bird netting can be particularly effective for large open areas such as roofs and loading bays. Design considerations include the type and material of the fixings utilized and the bird species requiring exclusion.

Mining Ponds

Miners will use chemical agents to extract minerals or metals from crushed rocks. These harmful chemical agents must be confined from volatiles, especially migratory species. In the United States EPA mandates that such cyanide ponds be covered at all times to prevent loss of wildlife.

Bird Scarer

A bird scarer is any one of a number devices designed to scare birds, usually employed by farmers to dissuade birds from eating recently planted arable crops.

They are also used on airfields to prevent birds accumulating near runways and causing a potential hazard to aircraft.

A traditional scarecrow.

Visual Scarers

Scarecrow

One of the oldest designs of bird scarer is the scarecrow in the shape of a human figure. The scarecrow idea has been built upon numerous times, and not all visual scare devices are shaped like humans. The "Flashman Birdscarer," Iridescent tape, "TerrorEyes" balloons, and other visual deterrents are all built on the idea of visually scaring birds. This method doesn't work so well with all species, considering that some species frequently perch on scarecrows.

Hawk Kite

A stationary modelled owl used as a bird scarer at Lysaker, Norway.

Many species of bird are also naturally afraid of predators such as birds of prey. "Hawk kites" are designed to fly from poles in the wind and hover above the field to be protected. They are shaped to match the silhouette of a bird of prey.

Helikites

The Helikite birdscarer is a lighter-than-air combination of a helium balloon and a kite. Helikites fly up to 200 ft in the air with or without wind. Although they do not look like hawks, they fly and hover high in the sky behaving like birds of prey. Helikites successfully exploit bird pests' instinctive fear of hawks and can reliably protect large areas of farmland.

Lasers

The use of lasers can be an effective method of bird scaring, although there is some evidence to suggest some birds are "laser-resistant". As the effectiveness of the laser decreases with increasing light levels, it is likely to be most effective at dawn and dusk. Although some lasers prove to be effective during daylight hours.

The method relies on birds being startled by the strong contrast between the ambient light and the laser beam. During low light conditions this technique is very selective and can be attuned to frequencies and wavelengths that individual bird species don't like, but at night the light beam is visible over a large distance and can cause widespread (non-species specific) disturbance.

Available on the market are manual operated laser torches or automated robots to move the laser beam automatically towards the birds.

Dead Birds

The use of model or actual dead birds is used to signal danger to others. Initially birds often approach the corpse but usually leave when they see the unnatural position of the bird. This approach has been frequently used in attempts to deter gulls from airports. Pheasant feed sacks often have an image of an owl with large eyes so that when empty they can be strung up to scare predators.

Balloons

An example of a visual bird-scare balloon

Balloons are an inexpensive deterrent. However, this method relies on the movement of balloons, which is something that birds can become used to. The addition of eye illustrations on the balloons has been shown to increase this method's effectiveness as it combats the birds' ability to adapt. Commercially available "scare-eye" balloons have holographic eyes that follow birds wherever they go. The long-term effectiveness of this method can be increased by periodically moving the placement of the scare devices.

In the United Kingdom the use of balloons is subject to approval from the Civil Aviation Authority, especially around airfields.

Auditory Scarers

Audible bird scarers use noise stimuli that makes birds uncomfortable. However, once birds realise these pose no real threats, they can easily become habituated to sounds that seemed initially frightening. If just being placed in situ and left, audible bird scarers can easily become ineffective bird control solutions, however when managed on an on-going basis or used as part of a greater bird deterrent system, sound methods can deliver quality results.

Sōzu (Shishi Odoshi)

One very old design is the Japanese sōzu, known metonymically as a shishi odoshi (although the term *shishi odoshi* properly refers to any method of scaring wild animals, including the Western scarecrow). Instead of using a visual method to distract pests, as the scarecrow does, it uses the sound of a heavy pipe repeatedly and rhythmically hitting a rock, using water as a timing device. The sōzu is also used in Japanese popular culture to denote inordinate amounts of wealth, combined with a traditional sensibility: by design, the shishi odoshi uses copious amounts of water, meaning either a very high water bill, or that it is situated on high value land with a stream or river running through it.

Propane Cannons

A typical propane gas gun bird scarer.

Towards the end of the 20th century, one of the most popular types of bird scarer used by farmers in Europe and America was the propane powered gas gun which produces a periodic loud explosion. The audible bang can reach volumes in excess of 150 decibels near the gun.

One of the problems with gas gun scarers is that the loud bangs are disruptive to people living on nearby properties. Also, birds adapt quickly to any sound that does not vary its magnitude, pitch or time interval. Propane cannons become ineffective after a short while.

Electronic Repellers

Sonic bird repellers are not effective; the birds quickly aclimate to them. Usually consisting of a central unit and several speakers, the system emits digitally recorded distress calls of birds, and, in some cases, calls of predators of the target species. Some emitters randomize pitch, magnitude, time interval, sound sequence and other factors in an attempt to prevent birds from getting used to them. Many of the sounds produced are regarded as annoying to people.

Ultrasonic Scarers

Ultrasonic devices are static sound-emitting bird deterrents, which, in theory, will annoy birds to keep them away from enclosed or semi-enclosed areas. Ultrasonic scarers are not harmful to birds, however there is debate around birds' ability to hear these frequencies at loud enough decibels. Birds are believed to have similar hearing to humans, with studies showing birds do not hear on an ultrasonic level, meaning that ultrasonic scarers often have little or no effect in deterring birds.

Cartridge Scarers

Cartridge scarers include a wide variety of noise-producing cartridges usually fired from rockets or rope bangers, or on aerodromes from modified pistols or shotguns, which produce a loud bang and emit flashes of light. They include shellcrackers, screamer shells and whistling projectiles, exploding projectiles, bird bangers and flares. Bird banger cartridges commonly use a low explosive known as flash powder.

Cartridges are projected from a shotgun with a range of 45–90 metres (148–295 ft), or pistols with a range of approximately 25 metres (82 ft), before exploding. Bird scaring cartridges can produce noise levels of up to 160 dB at varying ranges but in some countries both the cartridges and the gun require a firearms certificate.

Pyrotechnics have proved effective in dispersing birds at airports, landfill sites, agricultural crops and aquaculture facilities. At airports in the United Kingdom, shellcrackers fired from a modified pistol are the most common means of dispersing birds, as they allow the bird controller to have some directional control over birds in flight, so they can be steered away from runways.

However, as with all similar noises, there is a high risk of birds becoming used to any pyrotechnics or cartridge explosions.

Benign Acoustic Deterrence

In 2013, Dr. John Swaddle and Dr. Mark Hinders at the College of William and Mary created a new method of deterring birds using benign sounds projected by conventional and directional (parametric) speakers. The initial objectives of the technology were to displace problematic birds from airfields to reduce bird strike risks, minimize agricultural losses due to pest bird foraging, displace nuisance birds that cause extensive repair and chronic clean-up costs, and reduce bird mortality from flying into man-made structures. The sounds, referred to as a "Sonic Net," do not have to be loud and are a combination of wave forms - collectively called "colored noise" - forming non-constructive and constructive interference with how birds talk to each other.

Technically, the Sonic Nets technology is not a bird scarer, but discourages birds from flying into or spending time in the target area. The impact to the birds is similar to talking in a crowded room, and since they cannot understand each other they go somewhere else. Early tests at an aviary and initial field trials at a landfill and airfield indicate that the technology is effective and that birds do not habituate to the sound. The provisional and full patents were filed in 2013 and 2014 respectively, with further research and commercialization of the technology being ongoing.

Other

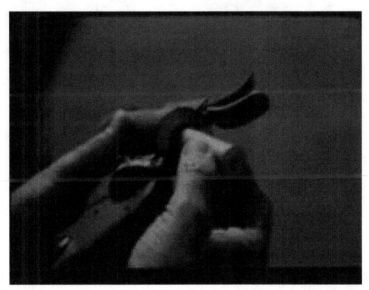

1986 U.S. Air Force video about methods to scare birds away from airfields

Dogs

The control of birds and other wildlife such as deer through harassment by trained border collies has been used at aerodromes, golf courses and agricultural land. The dogs represent an actual threat, and so elicit flight reactions. Habituation is unlikely as they can continually pursue and change their behaviour. Border collies are used as they are working dogs bred to herd animals and to avoid attack, and they respond well to whistle and verbal commands. A single border collie and its handler can keep an area of approximately 50 square kilometres (19.3 square miles, 4998.7 hectares, or 12,179.2 acres) free of larger birds and wildlife and although they are effective at deterring ground foraging birds such as waders and wildfowl, they are not so useful for species that spend most of their time flying or perching, such as raptors and swallows.

In 1999 Southwest Florida International Airport became the first commercial airport in the world to employ a border collie in an airfield wildlife control programme. After the use of the collie, numbers and species of birds on the airport declined and most birds that remained congregated in a drainage ditch away from the runway. The number of bird strikes dropped to zero compared to 13 for the same period the previous year. Several other airports and airbases have now started similar programmes.

At Dover Air Force Base, Delaware, bird strike damage to aircraft caused by birds has been reduced from an average of US $600,000 / year for the proceeding two years to US$24 000/year after the initiation of a bird control programme that included the use of border collies.

Predators

Harris's Hawk

Using predators as a natural bird deterrent has become a recommended form of controlling bird infestations. Specially selected species are trained to deal with working in un-natural environments with distractions and dangers they would not usually encounter.

The success of this method of bird control is based on the fact that many birds have a natural fear of falcons and hawks as predators, so their presence in the area encourages problem species to disperse. The natural reaction of most prey species is to form a flock and attempt to fly above the falcon. If this fails, they will attempt to fly for cover and leave the area.

Radio Controlled Aircraft

Radio-controlled model aircraft have been used to scare or 'haze' bird pests since the early 1980s, mainly over airfields, but have also been used over agricultural areas, fisheries and landfill sites. This method has been shown to be very effective and birds habituate more slowly to a treatment in which they are being actively hazed. At Whiteman Air Force Base, Missouri, balsa wood radio-controlled aircraft are one of the primary bird harassment methods used to keep the airfield clear of raptors and other large birds, and they have also proved effective at dispersing the base's redwing blackbird roost.

Fireworks

Fireworks can also be used as bird scarers, and some jurisdictions issue special licences for agricultural fireworks. This practice has been criticised as a loophole for the sale of consumer fireworks. Again, the loud bangs can also irritate people living on nearby properties.

Combination Scarers

These combine multiple deterrents, such as using a pop up scarecrow combined with a gas gun, which in turn activates the distress call of a bird. These combination scarers are often managed by computers and synchronised across an area via the use of radio links. This synchronisation becomes more effective if there is some kind of detection system involved such as bird detecting radar. One example of such a radar system which is cost effective on horticultural crops is the BirdDeter system.

References

- Francis Borgio J, Sahayaraj K and Alper Susurluk I (eds) . Microbial Insecticides: Principles and Applications, Nova Publishers, USA. 492pp. ISBN 978-1-61209-223-2

- Flint, Maria Louise & Dreistadt, Steve H. (1998). Clark, Jack K., ed. Natural Enemies Handbook: The Illustrated Guide to Biological Pest Control. University of California Press. ISBN 978-0-520-21801-7.

- Acorn, John (2007). Ladybugs of Alberta: Finding the Spots and Connecting the Dots. University of Alberta. p. 15. ISBN 978-0-88864-381-0.

- Kaya, Harry K. et al. (1993). "An Overview of Insect-Parasitic and Entomopathogenic Nematodes". In Bedding, R.A. Nematodes and the Biological Control of Insect Pests. CSIRO Publishing. ISBN 978-0-643-10591-1.

- CS1 maint: Uses authors parameter (link) CS1 maint: Uses editors parameter (link)Xuenong Xu (2004). Combined Releases of Predators for Biological Control of Spider Mites Tetranychus urticae Koch and Western Flower Thrips Frankliniella occidentalis (Pergande). Cuvillier Verlag. p. 37. ISBN 978-3-86537-197-3.

- Golden Gate Gardening: For more information on bird protection net / anti bird net refer birdproofingsolutions. in The Complete Guide to Year-Round Food Gardening (2002) ISBN 1-57061-136-X, p. 151

- "The Chinese Scientific Genius. Discoveries and inventions of an ancient civilization: Biological Pest Control" (PDF). The Courier. UNESCO: 24. October 1988. Retrieved 5 June 2016.

- Shapiro-Ilan, David I; Gaugler, Randy. "Biological Control. Nematodes (Rhabditida: Steinernematidae & Heterorhabditidae)". Cornell University. Retrieved 7 June 2016.

- "Conservation of Natural Enemies: Keeping Your "Livestock" Happy and Productive". University of Wisconsin. Retrieved 7 June 2016.

- Wilson, L. Ted; Pickett, Charles H.; Flaherty, Donald L.; Bates, Teresa A. "French prune trees: refuge for grape leafhopper parasite" (PDF). University of California Davis. Retrieved 7 June 2016.

Postharvest: Issues and Challenges

Harvest is the final stage of cropping and plays a vital role in determining the profit potential of the cultivated crop. Post-harvest refers to that stage when the grain or produce is harvested from the plant; the main concern at this stage is to reduce spoilage, wastage and to ensure that the harvest is protected from pests, disease and decay. This chapter details the methods used to increase the post-harvest shelf life of the produce and the causes of post-harvest losses of grains and vegetables.

Postharvest

In agriculture, postharvest handling is the stage of crop production immediately following harvest, including cooling, cleaning, sorting and packing. The instant a crop is removed from the ground, or separated from its parent plant, it begins to deteriorate. Postharvest treatment largely determines final quality, whether a crop is sold for fresh consumption, or used as an ingredient in a processed food product.

Drying and bagging of peanuts in Jiangxia District, Hubei, China

Goals

The most important goals of post-harvest handling are keeping the product cool, to avoid moisture loss and slow down undesirable chemical changes, and avoiding physical damage such as bruising, to delay spoilage. Sanitation is also an important factor, to reduce the possibility of pathogens that could be carried by fresh produce, for example, as residue from contaminated washing water.

Drying chili peppers. Milyanfan, Kyrgyzstan

After the field, post-harvest processing is usually continued in a packing house. This can be a simple shed, providing shade and running water, or a large-scale, sophisticated, mechanised facility, with conveyor belts, automated sorting and packing stations, walk-in coolers and the like. In mechanised harvesting, processing may also begin as part of the actual harvest process, with initial cleaning and sorting performed by the harvesting machinery.

Initial post-harvest storage conditions are critical to maintaining quality. Each crop has an optimum range of storage temperature and humidity. Also, certain crops cannot be effectively stored together, as unwanted chemical interactions can result. Various methods of high-speed cooling, and sophisticated refrigerated and atmosphere-controlled environments, are employed to prolong freshness, particularly in large-scale operations.

Regardless of the scale of harvest, from domestic garden to industrialised farm, the basic principles of post-harvest handling for most crops are the same: handle with care to avoid damage (cutting, crushing, bruising), cool immediately and maintain in cool conditions, and cull (remove damaged items).

Postharvest Shelf Life

Once harvested, vegetable and fruit are subject to the active process of senescence. Numerous biochemical processes continuously change the original composition of the crop until it becomes unmarketable. The period during which consumption is considered acceptable is defined as the time of "postharvest shelf life".

Postharvest shelf life is typically determined by objective methods that determine the overall appearance, taste, flavour, and texture of the commodity. These methods usually include a combination of sensorial, biochemical, mechanical, and colorimetric (optical) measurements. A recent study attempted (and failed) to discover a biochemical marker and fingerprint methods as indices for freshness.

Postharvest Physiology

Postharvest physiology is the scientific study of the physiology of living plant tissues after they have denied further nutrition by picking. It has direct applications to postharvest handling in establishing the storage and transport conditions that best prolong shelf life.

An example of the importance of the field to post-harvest handling is the discovery that ripening of fruit can be delayed, and thus their storage prolonged, by preventing fruit tissue respiration. This insight allowed scientists to bring to bear their knowledge of the fundamental principles and mechanisms of respiration, leading to post-harvest storage techniques such as cold storage, gaseous storage, and waxy skin coatings. Another well-known example is the finding that ripening may be brought on by treatment with ethylene.

Post-harvest Losses (Grains)

Grain silos in Australia

Grains may be lost in the pre-harvest, harvest and post-harvest stages. Pre-harvest losses occur before the process of harvesting begins, and may be due to insects, weeds and rusts. Harvest losses occur between the beginning and completion of harvesting, and are primarily caused by losses due to shattering. Post-harvest losses occur between harvest and the moment of human consumption. They include on-farm losses, such as when grain is threshed, winnowed and dried, as well as losses along the chain during transportation, storage and processing. Important in many developing countries, particularly in Africa, are on-farm losses during storage, when the grain is being stored for auto-consumption or while the farmer awaits a selling opportunity or a rise in prices.

Potential for Loss

There is potential for loss throughout the grain harvesting and agricultural marketing chains. During stripping of maize grain from the cob, known as shelling, losses can occur when mechanical shelling is not followed up by hand-stripping of the grains that are missed. Certain shellers can damage the grain, making insect penetration easier. For crops other than maize, threshing losses occur as a result of spillage, incomplete removal of the grain or by damage to grain during the threshing. They can also occur after threshing due to poor separation of grain from the chaff during cleaning or winnowing. Incomplete threshing usually occurs in regions with high labour costs, particularly at harvest time, when labour is too scarce and expensive to justify hand-stripping after an initial mechanical thresh. Certain mechanical threshers are designed only for dry grain.

A wet season's paddy harvest may clog the screens and grain will be lost. Cleaning is essential before milling. On the farm, cleaning is usually a combination of winnowing and removal by hand of heavier items such as stones. Losses can be low when the operation is done carefully but high with carelessness. With correct equipment, cleaning losses should be low in mills, but grain may be separated together with dirt or, alternatively, dirt may be carried forward into the milling stages. In drying, grain that is dried in yards or on roads, as is common in parts of Asia, may be partially consumed by birds and rodents. Wind, either natural or from passing vehicles in the case of road drying, can blow grain away.

Manual rice mill in Vietnam

The main cause of loss during drying is the cracking of grain kernels that are eaten whole, such as rice. Some grains may also be lost during the drying process. However, failure to dry crops adequately can lead to much higher levels of loss than poor-quality drying, and may result in the entire harvest becoming inedible. Adequate drying by farmers is essential if grains are to be stored on-farm and poorly dried grains for the market need to be sold quickly to enable the marketing-processing chain to carry out adequate drying before the grains become spoilt. With a high moisture content, grain is susceptible to mould, heating, discoloration and a variety of chemical changes. Ideally, most grains should be dried to acceptable levels within 2–3 days of harvest. One of the problems in assessing levels of post-harvest loss is in separating weight loss caused by the very necessary drying operations from weight loss caused by other, controllable, factors.

Milling to remove the outer coats from a grain may take place in one or more stages. For paddy rice considerable mechanical effort is needed to remove these layers. Any weakness in the kernel will be apparent at this stage. Even with grain in perfect condition, correctly set milling and polishing machinery is essential to yield high processing outturns. Complete separation of edible from less-desired products is always difficult to achieve but, even so, there are significant differences in milling efficiency. In the case of rice, milling outturns can vary from 60% or less to around 67%, depending on the efficiency of the mill. Even a 1% increase in yield of whole grain rice can thus result in huge increases in national food resources.

Grains are produced on a seasonal basis. In many places there is only one harvest a year. Thus most production of maize, wheat, rice, sorghum, millet, etc. must be held in storage for periods varying from a few days up to more than a year. Storage therefore plays a vital role in grain supply chains. For all grains, storage losses can be considerable but the greatest losses appear to be of maize, particularly in Africa. Losses in stored grain are determined by the interaction between the grain, the storage environment and a variety of organisms.

Contamination by moulds is mainly determined by the temperature of the grain and the availability of water and oxygen. Moulds can grow over a wide range of temperatures, but the rate of growth is lower with lower temperature and less water availability. The interaction between moisture and temperature is important. Maize, for example, can be stored for one year at a moisture level of 15% and a temperature of 15 °C. However, the same maize stored at 30 °C will be substantially damaged by moulds within three months. Insects and mites (arthropods) can, of course, make a significant contribution towards the deterioration of grain, through the physical damage and nutrient losses caused by their activity.

The Black Rat (Rattus rattus)

They can also influence mould colonisation as carriers of mould spores and because their faecal material can be utilised as a food source by moulds. In general, grain is not infested by insects below 17 °C whereas mite infestations can occur between 3 and 30 °C and above 12% moisture content. The metabolic activity of insects and mites causes an increase in both the moisture content and temperature of infested grain. Another important factor that can affect mould growth is the proportion of broken kernels. There are about 1,700 species of rodents in the world, but only a few species contribute significantly to post-harvest losses. Three species are found throughout the world: the house mouse (*Mus musculus*), the black rat (*Rattus rattus*) and the brown rat while a few other species are important in Africa and Asia.

Actual Loss

In Africa, post-harvest losses from harvest to market sale amount to around 10-20%. Approximately 40% of these losses occur during storage at the farm and market, 30% during processing (drying, threshing, and winnowing), 20% in transport from the field to the homestead/farm, and the remaining 10% during transport to market.

Loss Assessment Methods

An attempt should be made to approximate the magnitude of the value of losses before time is spent on trying to reduce them. If this value proves to be low, expenditure of appreciable resources on reducing losses may not be justified. However, despite efforts over the years to develop acceptable techniques for measuring grain losses, this remains an imperfect science. A particular problem with measurement is that grain does not follow a uniform sequence from producer to consumer. Harvested grain can be specially dried and treated for a family's consumption or for use as seed. Some of any harvest may be held for short-term storage, some more for long-term storage, and the rest may be sold either in one go or over a period of time, through a variety of different marketing channels. There are particular difficulties associated with accurately measuring on-farm storage losses over a long period when farmers are continually removing grain from stores to meet their own consumption needs. Further, the surplus generated by a farmer at any one harvest will dictate the quantity stored and the quantity sold, which, in turn, may influence loss levels. Given the lack of a consistent chain, care must be taken to avoid generalizing from particular measurements. *"Inordinately high- and low-loss situations must be put into perspective rather than giving them overemphasis as has been the case in some instances."*

The origin and justification of grain-loss estimates has thus never been particularly well- founded and attempts to measure losses suffer from the fact that it is an extremely complex and costly exercise to do well. To get round this problem the African Postharvest Losses Information System (APHLIS), was established in 2009. APHLIS generates weight loss data using an algorithm that refers to a postharvest loss profile (PLP) that is specific to the cereal crop, climate and scale of farming (smallholder or large scale) in question. The PLP is a set of loss figures, one for each link in the postharvest chain. Each PLP figure is the average of all those data available in the scientific literature for a particular crop (which include both quantitative weight loss figures and 'informed guestimates'), under a particular climate, and at a particular scale of agriculture. Given data on production and certain other relevant seasonal data, APHLIS can provide weight loss estimates for the provinces of many countries in Sub-Saharan Africa. The data are provided in tables and as interactive maps. A further important feature of APHLIS is that it provides a version of its loss calculator that can be downloaded from the website as an Excel file. Users can change default values in the spreadsheet and make calculations of losses at any desired geographical scale below the level of 'province'. With this calculator, users can go beyond estimation of losses at one link in the postharvest chain, e.g. just storage losses, which was the typical approach of the 1970s, and instead by substituting what figures they have for the default values in the PLP they can generate an estimate of cumulative losses from production, in other words they can see the changes in cereal grain supply that result from improving or deteriorating losses across the postharvest value chain. APHLIS thus provides data that are transparent in the way they are calculated, adjustable year by year according to circumstances, and upgradeable as more (reliable) data become available.

Attempts at Loss Reduction

There have been numerous attempts by donors, governments and technical assistance agencies over the years to reduce post-harvest losses in developing countries. Despite these efforts, losses are generally considered to remain high although, as noted, there are significant measurement difficulties. One problem is that while engineers have been successful in developing innovations in drying and storage these innovations are often not adopted by small farmers. This may be because farmers are not convinced of the benefits of using the technology. The costs may outweigh the perceived benefits and even if the benefits are significant the investment required from farmers may present them with a risk they are not prepared to take. Alternatively, the marketing chains may not reward farmers for introducing improvements. While good on-farm drying will lead to higher milling yields or reduced mycotoxin levels this means nothing to farmers unless they receive a premium for selling dry grains to traders and mills. This is often not the case.

Thus part of the problem with uptake may have been an overemphasis on technology, to the exclusion of socio-economic considerations. In the case of drying, it may be a more appropriate solution to strengthen the capacity of mills and traders to dry than attempt village-level improvements. There is thus a continual need to balance and blend technically ideal procedures and approaches with social, cultural, and political realities. Past on-farm storage interventions that have proved less than successful have included the promotion of costly driers in W. Africa that fell victim to termites when made with local wood or bamboo and were too expensive when constructed with sawn wood. In the 1980s, there was considerable enthusiasm for the introduction of ferro-cement and brick bins throughout Africa, but these were often found to be too complicated for farmers to construct, and too costly. Small Breeze block silos also experienced construction difficulties and were found to be not economically feasible. Storage cribs made of wood and chicken-wire were introduced by donors but rejected by farmers because sides made of chicken wire showed others the size of each farmer's harvest.

More positive achievements have been recorded in the Central African Republic, using a simple 1-tonne capacity structure that was found by farmers to be easy to construct and proved popular even without donor subsidies. Considerable success has been reportedly achieved with metal bins over the last 20 years in Central America and metal bins have been widely used for grain storage in Swaziland for half a century, drawing on the availability of local entrepreneurs who had been supplying metal water tanks. Replication of this success in other parts of Africa is very much in the pilot stage. Difficulties include the lack of local craftsmen to fabricate the bins; the need for grain stored in such bins to be dried to 14 °C, and problems with carrying out the necessary fumigation. Small-scale bins for use inside the home appear to be having more success than larger bins for outside use. A relatively new development is hermetically sealed bags, which appear to offer good possibilities to store a variety of quantities, although further socio-economic evaluation is still required. The Purdue Improved Cowpea Storage (PICS) bags are hermetically sealed bags that allow small-scale farmers/users to store cowpea without any use of chemicals.

Post-harvest Losses (Vegetables)

The post-harvest sector includes all points in the value chain from production in the field to the food being placed on a plate for consumption. Postharvest activities include harvesting, handling, storage, processing, packaging, transportation and marketing.

Losses of horticultural produce are a major problem in the post-harvest chain. They can be caused by a wide variety of factors, ranging from growing conditions to handling at retail level. Not only are losses clearly a waste of food, but they also represent a similar waste of human effort, farm inputs, livelihoods, investments and scarce resources such as water. Post-harvest losses for horticultural produce are, however, difficult to measure. In some cases everything harvested by a farmer may end up being sold to consumers. In others, losses or waste may be considerable. Occasionally, losses may be 100%, for example when there is a price collapse and it would cost the farmer more to harvest and market the produce than to plough it back into the ground. Use of average loss figures is thus often misleading. There can be losses in quality, as measured both by the price obtained and the nutritional value, as well as in quantity.

Discarded tomatoes on a compost heap at nurseries in the UK

On-farm Causes of Loss

There are numerous factors affecting post-harvest losses, from the soil in which the crop is grown to the handling of produce when it reaches the shop. Pre-harvest production practices may seriously affect post-harvest returns. Plants need a continuous supply of water for photosynthesis and transpiration. Damage can be caused by too much rain or irrigation, which can lead to decay; by too little water; and by irregular water supply, which can, for example, lead to growth cracks. Lack of plant food can affect the quality of fresh produce, causing stunted growth or discoloration of leaves, abnormal ripening and a range of other factors. Too much fertilizer can harm the development and post-harvest condition of produce. Good crop husbandry is important for reducing losses. Weeds compete with crops for nutrients and soil moisture. Decaying plant residues in the field are also a major loss factor.

Causes of Loss After Harvest

Fruits and vegetables are living parts of plant and contain 65 to 95 percent water. When food and water reserves are exhausted, produce dies and decays. Anything that increases the rate at which a product's food and water reserves are used up increases the likelihood of losses. Increases in normal physiological changes can be caused by high temperature, low atmospheric humidity and physical injury. Such injury often results from careless handling, causing internal bruising, splitting and skin breaks, thus rapidly increasing water loss.

Respiration is a continuing process in a plant and cannot be stopped without damage to the growing plant or harvested produce. It uses stored starch or sugar and stops when reserves of these are exhausted, leading to ageing. Respiration depends on a good air supply. When the air supply is restricted fermentation instead of respiration can occur. Poor ventilation of produce also leads to the accumulation of carbon dioxide. When the concentration of carbon dioxide increases it will quickly ruin produce.

Fresh produce continues to lose water after harvest. Water loss causes shrinkage and loss of weight. The rate at which water is lost varies according to the product. Leafy vegetables lose water quickly because they have a thin skin with many pores. Potatoes, on the other hand, have a thick skin with few pores. But whatever the product, to extend shelf or storage life the rate of water loss must be minimal. The most significant factor is the ratio of the surface area of the fruit or vegetable to its volume. The greater the ratio the more rapid will be the loss of water. The rate of loss is related to the difference between the water vapour pressure inside the produce and in the air. Produce must therefore be kept in a moist atmosphere.

Diseases caused by fungi and bacteria cause losses but virus diseases, common in growing crops, are not a major post-harvest problem. Deep penetration of decay makes infected produce unusable. This is often the result of infection of the produce in the field before harvest. Quality loss occurs when the disease affects only the surface. Skin blemishes may lower the sale price but do not render a fruit or vegetable inedible. Fungal and bacterial diseases are spread by microscopic spores, which are distributed in the air and soil and via decaying plant material. Infection after harvest can occur at any time. It is usually the result of harvesting or handling injuries.

Ripening occurs when a fruit is mature. Ripeness is followed by senescence and breakdown of the fruit. The category "fruit" refers also to products such as aubergine, sweet pepper and tomato. Non-climacteric fruit only ripen while still attached to the parent plant. Their eating quality suffers if they are harvested before fully ripe as their sugar and acid content does not increase further. Examples are citrus, grapes and pineapple. Early harvesting is often carried out for export shipments to minimise loss during transport, but a consequence of this is that the flavour suffers. Climacteric fruit are those that can be harvested when mature but before ripening has begun. These include banana, melon, papaya, and tomato. In commercial fruit marketing the rate of ripening is controlled artificially, thus enabling transport and distribution to be carefully planned. Ethylene gas is produced in most plant tissues and is important in starting off the ripening process. It can be used commercially for the ripening of climacteric fruits. However, natural ethylene produced by fruits can lead to in- storage losses. For example, ethylene destroys the green colour of plants. Leafy vegetables will be damaged if stored with ripening fruit. Ethylene production is increased when fruits are injured or decaying and this can cause early ripening of climacteric fruit during transport.

Damage in the Marketing Chain

Tomato harvesting in Portugal

Fruits and vegetables are very susceptible to mechanical injury. This can occur at any stage of the marketing chain and can result from poor harvesting practices such as the use of dirty cutting knives; unsuitable containers used at harvest time or during the marketing process, e.g. containers that can be easily squashed or have splintered wood, sharp edges or poor nailing; overpacking or underpacking of containers; and careless handling of containers. Resultant damage can include splitting of fruits, internal bruising, superficial grazing, and crushing of soft produce. Poor handling can thus result in development of entry points for moulds and bacteria, increased water loss, and an increased respiration rate.

Produce can be damaged when exposed to extremes of temperature. Levels of tolerance to low temperatures are importance when cool storage is envisaged. All produce will freeze at temperatures between 0 and -2 degrees Celsius. Although a few commodities are tolerant of slight freezing, bad temperature control in storage can lead to significant losses.

Some fruits and vegetables are also susceptible to contaminants introduced after harvest by use of contaminated field boxes; dirty water used for washing produce before packing; decaying, rejected produce lying around packing houses; and unhealthy produce contaminating healthy produce in the same packages.

Losses directly attributed to transport can be high, particularly in developing countries. Damage occurs as a result of careless handling of packed produce during loading and unloading; vibration (shaking) of the vehicle, especially on bad roads; and poor stowage, with packages often squeezed into the vehicle in order to maximise revenue for the transporters. Overheating leads to decay, and increases the rate of water loss. In transport it can result from using closed vehicles with no ventilation; stacking patterns that block the movement of air; and using vehicles that provide no protection from the sun. Breakdowns of vehicles can be a significant cause of losses in some countries, as perishable produce can be left exposed to the sun for a day or more while repairs are carried out.

At the retail marketing stage losses can be significant, particularly in poorer countries. Poor-quality markets often provide little protection for the produce against the elements, leading to rapid produce deterioration. Sorting of produce to separate the saleable from the unsaleable can result in high percentages being discarded, and there can be high weight loss from the trimming of leafy vegetables. Arrival of fresh supplies in a market may lead to some existing, older stock being discarded, or sold at very low prices.

Avoiding Loss

Losses can be avoided by following good practices as indicated above. There is also a wide range of post-harvest technologies that can be adopted to improve losses throughout the process of pre-harvest, harvest, cooling, temporary storage, transport, handling and market distribution. Recommended technologies vary depending on the type of loss experienced. However, all interventions must meet the principle of cost-effectiveness. In theory it should be possible to reduce losses substantially but in practice this may be prohibitively expensive. Especially for small farms, for which it is essential to reduce losses, it is difficult to afford expensive and work-intensive technologies.

Assessing Losses

There are no reliable methods for evaluating post-harvest losses of fresh produce. Any assessment can only refer to a particular value chain on a particular occasion and, even then, it is difficult to account for quality loss or to differentiate between unavoidable moisture loss and losses due to poor post-harvest handling and other factors described above. Accurate records of losses at various stages of the marketing chain are rarely kept, particularly in tropical countries where losses can be highest, making reliable assessment of the potential cost-effectiveness of interventions at different stages of the chain virtually impossible. The lack of such information may lead to misplaced interventions by governments and donors.

Permissions

All chapters in this book are published with permission under the Creative Commons Attribution Share Alike License or equivalent. Every chapter published in this book has been scrutinized by our experts. Their significance has been extensively debated. The topics covered herein carry significant information for a comprehensive understanding. They may even be implemented as practical applications or may be referred to as a beginning point for further studies.

We would like to thank the editorial team for lending their expertise to make the book truly unique. They have played a crucial role in the development of this book. Without their invaluable contributions this book wouldn't have been possible. They have made vital efforts to compile up to date information on the varied aspects of this subject to make this book a valuable addition to the collection of many professionals and students.

This book was conceptualized with the vision of imparting up-to-date and integrated information in this field. To ensure the same, a matchless editorial board was set up. Every individual on the board went through rigorous rounds of assessment to prove their worth. After which they invested a large part of their time researching and compiling the most relevant data for our readers.

The editorial board has been involved in producing this book since its inception. They have spent rigorous hours researching and exploring the diverse topics which have resulted in the successful publishing of this book. They have passed on their knowledge of decades through this book. To expedite this challenging task, the publisher supported the team at every step. A small team of assistant editors was also appointed to further simplify the editing procedure and attain best results for the readers.

Apart from the editorial board, the designing team has also invested a significant amount of their time in understanding the subject and creating the most relevant covers. They scrutinized every image to scout for the most suitable representation of the subject and create an appropriate cover for the book.

The publishing team has been an ardent support to the editorial, designing and production team. Their endless efforts to recruit the best for this project, has resulted in the accomplishment of this book. They are a veteran in the field of academics and their pool of knowledge is as vast as their experience in printing. Their expertise and guidance has proved useful at every step. Their uncompromising quality standards have made this book an exceptional effort. Their encouragement from time to time has been an inspiration for everyone.

The publisher and the editorial board hope that this book will prove to be a valuable piece of knowledge for students, practitioners and scholars across the globe.

Index

www.ingramcontent.com/pod-product-compliance
Lightning Source LLC
Jackson TN
JSHW052210130125
77033JS00004B/224

* 9 7 8 1 6 3 5 4 9 0 7 9 4 *